The Piano, You Can'

GW01315919

An Alphabetical Exploration and Celebration of the Humble Pianoforte

Published by Lewarne Publishing 2024

www.lewarne-publishing.co.uk

Printed and bound by Catford Print, London

For Christine

Other books by Steven Harris include:
Painting in Earnest (Life of painter Ernest Stafford Carlos)
Meet the Baden-Powells (Lord Baden-Powell and Family)
The Man from Harrods (Piano Tuner Memoirs)

For over 300 years the pianoforte has never been very far away,
it's still very much out and about...

CONTENTS

Y YAYOI KUSAMA'S YELLOW DOT PIANO 194; YAMAHA 195

Z ZOOS 196; ZHU ZAIYU 196; ZEBRA 196-197; (for Zumpe, prodigious maker of square pianos, see page 91)

Picture credits: pages 198 to 200

ACKNOWLEDGEMENTS

Grateful thanks are here extended to the people and organizations who helped the author with information and support, they include: the archive departments of Bechstein, Blüthner, Bösendorfer, and Steinway; Matt Ash and the Music Industries Association; Colin Batt Removals; Colonel Blashford-Snell CBE; the British Library; the Cobbe Collection Trust, Hatchlands Park; Barrie Heaton FIMIT, FABPT; Finchcocks Musical Instrument Collection and Study Centre; Ken Forrest, former president of the Pianoforte Tuners' Association; the Glenn Gould Foundation; Stefan Gritzka; the Institute of Musical Instrument Technology; Dr Alastair Laurence; London Metropolitan Archives; the National Trust; Lori Norriss; Rachel Peat, assistant curator at the Royal Collection Trust; the Royal College of Music library and archives; the Scott Polar Research Centre, Cambridge; Sir Richard Stilgoe; Uniplex (UK) Ltd; the late Norman Wattam; Chris Winter; Yamaha (Europe) Music Company.

Front cover image courtesy of Yamaha; rear cover images courtesy of Alamy (Erard) and Edelweiss piano company.

The trail-blazing C. Bechstein Piano Centre, launched in Manchester in 2023.

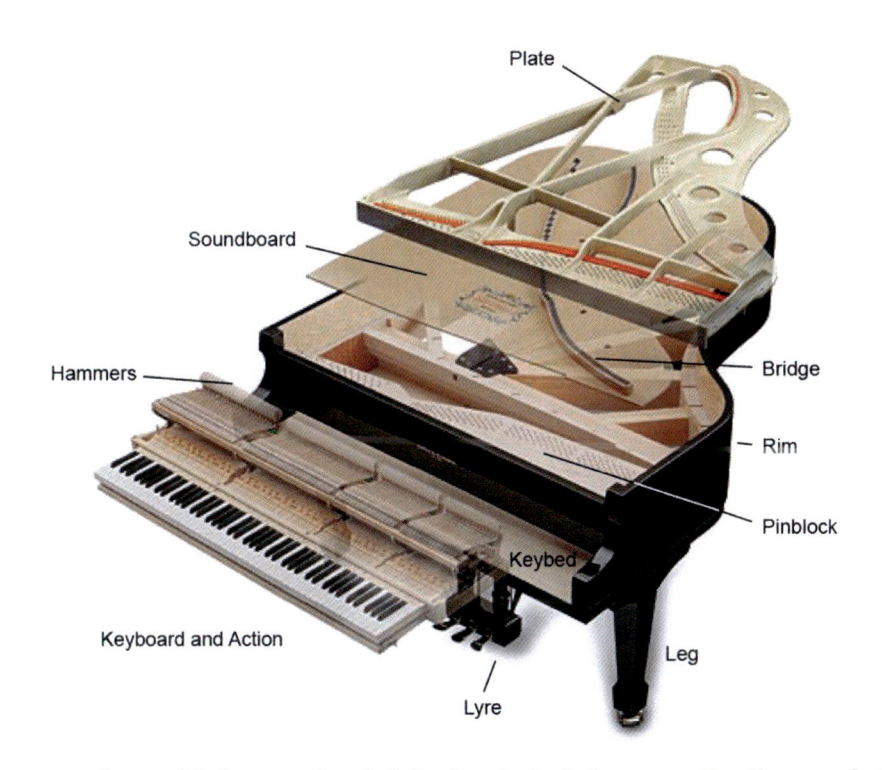

Other than here, this is not a book full of technical diagrams, the theory of piano construction, and jargon but the above and below images of the workings of a grand and upright piano provide basic nomenclature and a visual prompt (note: for the grand piano image, the strings are missing, as are the dampers that rest on the strings and stop them from vibrating until a note is played or the sustaining pedal is used). The playing mechanism of the piano is known as the 'action'. On a grand, the action is screwed as one unit on to the keyboard (which can be slid in and out of the piano); on an upright, the action can be lifted in and out of the piano but is not bolted on to the keyboard as the keyboard is normally built permanently into the piano's case/body.

UPRIGHT PIANO ACTION

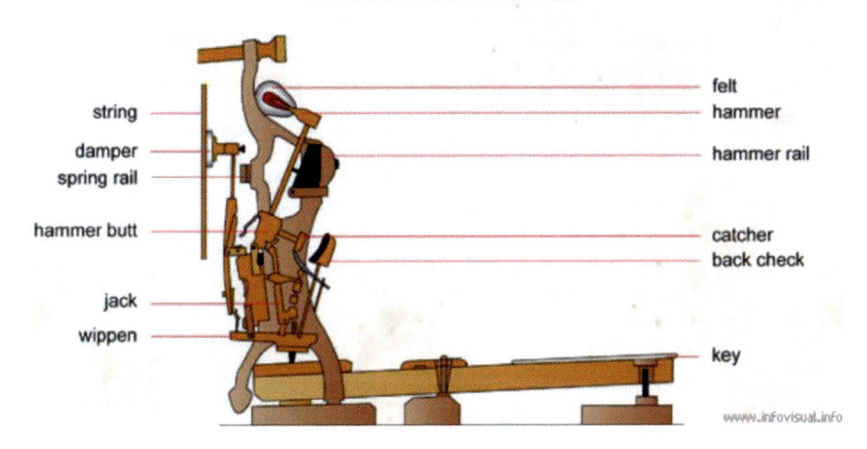

INTRODUCTION

The piano is dead. No one buys pianos anymore, no one plays them. What a relic they are, bits of clapped out wood and yellowing ivory; most have been either dumped or smashed to pieces.

Myths and legends can be sad, wonderful, entertaining and often untrue. The acoustic ('real') piano is still very much alive and playing. Ancient ones are sometimes scrapped or recycled into handsome bookcases, garden features and even pieces of art, but many people do still use and buy pianos and they are often more prominent than in latter days – they are also widely seen and played in restaurants and at many train stations and other public places. Music colleges and academies throughout the land have countless acoustic pianos (most had lost count when I enquired though the Royal College of Music confirmed they had at least one hundred and sixty-eight).

Those who buy a brand new one will become owners of an instrument that could last a lifetime, many have never been made better. True, the piano – invented long before such things as telephones, radios and cars (even bicycles) – has to compete with many modern technological devices and other forms of entertainment, yet thousands of unknown users close a door, sit quietly down and use them to regain moments of pleasure, tranquillity and sanity in a hectic world. The piano, it could be said, is a far better antidepressant than the pills once rather too eagerly prescribed by doctors. Indeed, let's not knock the piano, it needs celebrating!

The acoustic piano has its digital rival (or relative, imposter, ally…) but these instruments, too, are still 'pianos'. And while sales of acoustic pianos in this country these days may be steadier than the once mega-volume numbers, pianos sold today are far less likely to be purchased as symbols of culture or furniture to aesthetically complement the layout of a room. The piano today is nearly always *wanted* primarily as a musical instrument, not a mere adornment or status symbol. Worldwide sales of pianos are high. Never more so in China, where the piano was once another kind of symbol – political – and outlawed under the Cultural Revolution. No wonder, then, that Chinese pianist Zhu Xiao-Mei, perhaps even more so than almost every other owner of a piano, should consider her small and rather ordinary upright piano to be almost an intimate friend. It had to be hidden in the home, and at another time moved somewhere more secret. Finally, hoping to be reunited with it on a desolate railway station, she wrote:

My heart started beating faster. My piano had arrived! In a few minutes I was going to see it again. I ran in search of the platform and found myself in a sort of wasteland, where I could see, in the distance, a small, black, shapeless shabby object … Clearly it had travelled in the coal compartment, but it had arrived, and was in one piece.

I leaned against it, and I spoke to it softly: "I'll never leave you again, I swear. Never!"

Today, China has more piano factories than any other country and produces thousands of pianos a year (many of which are bought by Westerners but, interestingly, there is a niche market in countries like China and Japan, where some local piano buyers lap up any available Western-made piano, be it a Broadwood or Barnes, Chappell, Challen or Cavendish). And although it is a huge part of Asian culture today, after making its tentative appearance among royal households first and then filtering down to the gentry and other classes, the pianoforte (or simply piano) both in word and object became known by practically everybody. Dickens has a character wanting to hide someone in a disembowelled piano (calling it a *pianner forty*) in *The Pickwick Papers*; a 'crabbed old piano tuner' (actor) walked on to the St James' stage in 1843, and pianos or piano tuners have featured in poetry, art, plays, television and film – also in ballet and opera – right up to and beyond Daniel Mason's 2002 novel, *The Piano Tuner*. The piano is both nostalgia and current, still seen and heard at concerts large and small, still played competitively but also for wonderful pleasure at home and abroad.

Described as an 'intriguing mixture of pedals, pins, and paradox' by pianist Glenn Gould, what a very special and unique instrument the piano is. No wonder it has been voted the world's most popular musical instrument, for it is a harp in a box, a complete orchestra, an amazing machine with around 12,000 individual parts. For tonal range, the piano can outmatch any instrument of the orchestra; it is also a keyboard, string and percussion instrument that evolved from a rather primitive harpsichord with hammers, to being a must-have household item, piece of art, status symbol, popular wedding gift and phenomenal concert instrument capable of, in the right hands, enthralling audiences as a solo instrument, moving others to tears or being used as accompaniment and in beautiful collaboration with singers and ensemble groups or alternatively holding its own with a full orchestra.

The piano is such a versatile and resilient instrument. Even when maintained poorly or abused, most refuse to die, still having the capacity for people to knock a good tune out of them. And perhaps a subtitle for this book should be *The Bouncing Piano*, for although over its long history its usage has declined at times or people have written it off as other new and wonderful (allegedly) forms of entertainment have come to the fore, sure enough the piano sooner or later reemerges in popularity. The whining wireless, sinister cinema and tedious television, if you'll forgive my bias, were a threat to the piano, as were other attractions later on (and continue to be to this day), leading many to forecast the instrument's extinction, yet it has always come back into favour and, three centuries or so after its invention, it continues to bounce along and give bountiful pleasure to billions around the world.

There have been many books on the piano, some relating its history, others rather more detailed – if somewhat dry – and with technical diagrams. The internet is also awash with information and misinformation about pianos. Be warned, too, for in the world of pianos, both past and recent piano makers, industry 'experts' and authors from different countries have made conflicting claims about who invented what and when (correct facts can be hard to come by). I humbly asked myself, however, if I could write something that would be a bit different – hopefully a combination of pleasing pictorial evidence, being broadly informative but also with occasional humour and quirkiness. I decided to set myself a challenge of producing a less heavyweight book based on the above but linking the piano (and so by definition pianists and even anything loosely to do with the word 'piano' itself) to at least one letter of the entire alphabet.

So this is the kind of book that can be read as a general reference source, read from the beginning or simply dipped into at random. Please feel free to open the 'lid' and play the first note you come across or start on the piano's first note, A, and read the entries for A before ascending up the scale to B, then C (my 'scale' will get you all the way to Z). No matter your preference, let the sightreading begin...

Salvador Dalí's bronze *Surrealist Piano*. The inanimate object of a piano is animated by the addition of dancers' legs and a gold ballet dancer. Measuring 60cm high from the base, it was priced at between £5000 and £7000 by one dealer. More can be read on art with a piano theme from page 15.

A

ABBEY ROAD STUDIOS; AVIATION; ALIQUOT; AMERICAN VOCABULARY; ART (PIANO THEMED)

The Abbey Road Studios in London's St. John's Wood area is where The Beatles made some of their famous recordings; in fact, the studios were earlier named EMI Studios, but officially took on the name Abbey Road after the release of The Beatles' album of the same name. Used by the great and good, in all genres of music, it was one of the earliest places where Sir Elton John began learning his craft in the music industry. As an unknown session musician, among other things he worked alongside the rather staid but highly professional Mike Sammes Singers, who sang backing tracks for many top performers, including Tom Jones and The Beatles. (The studios' hanging sound baffles in those times were apparently filled with dried seaweed.)

The Steinway vertegrand upright on which Paul McCartney played and recorded *Lady Madonna* and other hits is still there (Sir Paul politely ignored the Don't Touch sign when he visited the studios more recently). Originally a Georgian town house with nine bedrooms and large garden, Abbey Road Studios was opened as a recording studio by composer Sir Edward Elgar in 1931 (which included a performance of his *Pomp and Circumstance Marches*). Acclaimed as the oldest purpose-built recording studio in the world, anyone can record there. Smallest studio daily rates start at around £600 though additional charges may crank fees up into the thousands.

Right: Today a listed building, it was built in 1831 and acquired by the Gramophone Company (later becoming EMI) in 1929 for £16,500.

An unlikely rising star who shared studios with The Beatles was a pianist in her forties. She was known simply as Mrs Mills.

East Londoner Gladys Jordan became the famous Mrs Mills, known for her rather unique honky-tonk sing-along performances using a stride piano style – a time well before karaoke (which translates as 'empty orchestra'). She made her first television appearance on the *Billy Cotton Show* in 1961. Mrs Mills went on to record around forty LPs and appeared on numerous radio and television shows. When she first started recording, however, not wanting to risk her day job as a superintendent in a typing pool, she insisted on coming in and making the recordings during her lunchbreaks. The vintage 1905 Steinway vertegrand upright piano used by The Beatles and other artistes, eventually became known as the Mrs Mills piano. It was deliberately put out of tune by having one string on most of the notes detuned/flattened slightly and the hammer felts specially treated to get a more jangly, chorus effect (it had been purchased by Abbey Road Studios in 1953 for £404).

AVIATION

Having a piano on a cruise ship was and still is very common. In terms of aviation, most leading piano manufacturers halted production during war periods and made items to help the war effort (as the Chicago employee in the photo would soon do), for example parts for aeroplanes. While the female staff at Broadwood Pianos made ammunition boxes during the First World War, male staff made aircraft propellors and biplanes which were held together using piano wire (an unfortunate use for piano wire was known to be used on captured SOS agents in the Second World War: for strangulation and execution). The Bentley piano company based in the Cotswolds, similarly, produced parts for the Gloster Meteor jet, the first British jet fighter and the allies' only jet aircraft to achieve combat operations during the Second World War.

Although manufacturers of British pianos and parts virtually all fell by the wayside as the decades crept towards the end of the twentieth century, there is one company that managed to reinvent itself. Gamble piano actions were made for thousands of pianos. Although the firm ceased to trade under this name, in 1958 they grew into ENL and continue to make a range of aircraft components for the aviation industry (still with a Gamble as MD).

With the rise and popularity of luxury air travel in the 1970s, it wasn't unheard of for these planes to have a piano bar. Nonetheless, the Blüthner piano firm much earlier got a grand piano airborne by supplying a specially designed lightweight aluminium grand for the Hindenburg airship in 1936. The piano was said to have had a large and full tone. On the Hindenburg's maiden voyage, the Dresden pianist Franz Wagner played works by Chopin, Beethoven and Brahms. The airship (shown below), which had capacity for 50 to 70 passengers, made a total of 63 flights and much improved on Atlantic crossings. Cunard liners took around five days, the airship could do it comfortably in two. Fortunately, the Blüthner grand was later decommissioned, as the Hindenburg met its famous demise in May 1937, when it crashed while attempting to land. Bursting into flames, it took only a minute or so to be completely destroyed.

The story of the Hindenburg Blüthner was quite well publicized at the time, though information in the archives of the Bechstein piano company states they also supplied one of their grands for a Graz Zeppelin flight.

The first ever 'concert in the air' was broadcast live by 63 radio stations from around the world.

Being positioned in a not particularly large lounge on the airship, and with weight being an important factor, the Blüthner grand was specially designed to be lighter than the average baby grand, being 356 lbs as opposed to an average 500 to 600 lbs (227 to 272 kg). It also had thinner strings, and the tubular legs and lyre, along with the frame, were made out of a lighter metal: duralumin.

ALIQUOT

'Aliquot' is a rather odd sounding word, but it applies particularly to Blüthner grand pianos. Apart from the bass range, pianos have trichords – three strings to each note which have to sound identical – and Blüthner devised a new system in 1873 which used an additional fourth string to each note in the top three octaves. Placed in a raised position, the extra aliquot strings are not actually struck by the hammers. Tuned an octave higher than their note's counterpart, they are supposed to vibrate sympathetically and help produce an enriched tone from the piano.

AMERICAN VOCABULARY

In the piano world, American English can sometimes lead to confusion (and 'tuner' always seems to be pronounced 'tooner'). The piano's playing mechanism, which includes such things as the hammers, springs and small wooden levers, is universally known as the 'action', but other piano vocabulary is not the same. The piano tuner's tuning lever (or 'crank') is known as a tuning hammer (or the ghastly term 'wrench'); 'hammer' perhaps originates from the first tuning tools being T-shaped and having a blunt end on the handle for hammering-in tuning pins. The robust wooden section where the piano's tuning pins are hammered into, there again, is called in English the wrest plank, and in American the pinblock. The piano's iron frame is a harp or plate, piano stools are benches, upright pianos can be called verticals, the fall (lid over the keyboard) is the fallboard, while the music desk/rest is often called the music rack. Lastly, the rubber or felt wedges used by tuners to damp/mute certain strings are called mutes.

ART (PIANO THEMED)

The piano has captured the artist's gaze, hand and brush for almost as long as pianos have been around. You will see examples of pianos and pianists depicted in art elsewhere in this book, but what follows here are examples of the piano (or pianists) being depicted in a variety of artistic forms. The collection is deliberately wide-ranging; it is hoped the gallery here may whet the reader's appetite and strike a chord... (In 1936, the Kemble piano firm, incidentally, actually produced an attractive Cubist upright piano with chromium fittings and electric lights.)

Above: Work by Giovanni Boldini. He lived for a time in London before settling in Paris; he was once dubbed the Master of Swish because of his flowing style of painting.

Left: Modern artwork that featured in the *Piano Builder*; also, below right is a contemporary jazz poster.

Below left: By Salvador Dalí.

Above: Picasso's *Le Piano* (Velazquez).

Left: Van Gogh's *Marguerite Gachet au Piano*. Below: Famous for *The Scream*, this is Edvard Munch's *At the Grand Piano.*

Left: Paderewski (*Vanity Fair* 1899); above, a caricature of Franz Liszt (*La Vie Parisienne* 1886).

Left: Painted by Dalí in 1964, the Hamburg model A became known as the *Mozart Steinway*.

Below: In 1944 Dalí painted his *Sentimental Colloquy* (oil and ink on canvas). The design served as a backdrop for a ballet (described as a surrealist extravaganza) loosely based on one of Verlaine's poems and was staged in New York. It was said Dalí's piano and cyclists with rocks on their heads parodied Spain's provincial customs and the bourgeois culture of an elite public.

Left: German artist, Rebecca Horn's *Hanging Piano*, also titled *Concert for Anarchy*, which was exhibited (hanging from a high ceiling) in the Tate Gallery.

The lid was often closed but would catch visitors out by automatically opening (and the keys bursting out) at certain intervals. Thinking of machines, she said:
I like my machines to tire... They are more than objects. These are not cars or washing machines. They rest, they reflect, they wait.

Left and below: Chiharu Shiota's piano art piece in Yorkshire Sculpture Park (near Wakefield).

Overleaf: Work by New Zealand sculptor Michael Parekōwhai. Its title is the same as, and has a connection with, Keats' poem *On First Looking into Chapman's Homer.*

The talented sculptor (born 1968 and of Māori and Pākehā descent) has other exhibits in galleries around the world. The fur seal and piano sculpture, right, is titled *The Horn of Africa*.

Parekōwhai has used at least twelve pianos in his work. A 1920s Steinway grand which he transformed into a 'Maori Steinway' by his unique carvings on the case, came to the Oceania Exhibition in 2018 at London's Royal Academy of Arts. The reconditioned grand had originally belonged to Hungarian concert pianist Lili Kraus (who spent part of the second world war in a Japanese prisoner of war camp). A model D, the Steinway's home is normally at the Museum of New Zealand in Wellington but has been sent on numerous tours. The sculptor is insistent that it should be played where ever it is exhibited.

Below: An elaborate art-case Steinway piano, themed from Mussorgsky's *Pictures at an Exhibition*. It was designed, painted, unveiled and played by Steinway artist Paul Wyse. In 2017 it was priced at a snip of just $2.5 million.

The piano took four years to complete.

B

BECHSTEIN; BURNING PIANOS

It is rather coincidental that three of the major (and still functioning today) piano builders were all founded in the same year of 1853: Bechstein, Steinway and Blüthner (only the latter company has any family members still involved with the business). Another major player then and today, Bösendorfer, actually had a quarter of a century head start, having been founded in 1828. At a concert ten years later in Vienna, Liszt played a Bösendorfer piano and a firm friendship grew between pianist and company from then on.

Having arrived in Berlin around the year 1847, Carl Bechstein (shown on page 20) worked with piano maker Gottfried Perau. Bechstein resigned a year later and travelled to London then Paris to work with two piano makers of high repute. Although he did not work for the renowned Erard firm, he was also influenced by the design and high standard of work produced by this French maker (the highly successful Erard in fact would have piano factories in both Paris and London). After his sojourn abroad, Bechstein returned to Berlin in 1852 somewhat inspired and was appointed manager of Perau's factory. The following year, nonetheless, while working for Perau, he was allowed to build his own pianos and sign his name on them, though officially the foundation of his company did not start until 1856.

Carl Bechstein had had musical training since a youngster (learning piano, violin and cello); as a young adult he mixed in well-connected circles where he came across professional musicians and well-to-do patrons, one example being the pianist Hans von Bülow, who had studied with Liszt. Bülow would later marry Liszt's daughter (but who would later marry the composer Richard Wagner). In marketing terms, it was fortunate for Bechstein that the well-known Bülow was happy to show his keen approval of the Bechstein piano. Lizst, too, was happy to purchase a Bechstein for his own use.

At the Great London Exposition in 1862, a jury described the Bechstein as having a 'freshness and freedom of tone'. Wagner, similarly, wrote a letter after receiving a Bechstein, describing it as having a 'chrystaline and delightful voice' (though the positive letter might have been of the 'obligatory' kind, sent by those fortunate enough to have been provided with expensive instruments pro gratis).

The Bechstein firm would suffer a serious fire, but this seemed to be common form for piano makers. The English piano firm, Collard & Collard, was burnt to the ground twice, in 1807 and again in 1851 at the new Camden Town site. The Kirkman piano factory suffered the same fate in 1853, as did the Broadwood factory three years later. Erard in Paris lost around 1,500 pianos in a fire of 1888, while the Steinway Hamburg factory suffered a fire the following year. Even much later, the Bentley piano firm (Stroud company) had fires in 1938 and 1989. Fortunately, the Bechstein firm got off fairly lightly and recovered from the fire.

For a time, Carl Bechstein had to contend with dishonest imitators trying to cash-in on his growing reputation – a complimentary nickname around 1890 was 'The Prussian Erard'. Lesser firms were producing 'Ecksteins' and even 'Becksteins', forcing Bechstein to take legal action.

Perhaps the company benefitted from Queen Victoria's German link via Albert, her husband. In 1881, Bechstein pianos were supplied to Buckingham Palace and other royal residences. Soon after, in 1885 the first Bechstein showroom was opened in London; it proved to be highly successful. Four years later, they moved into London's Wigmore Street (number 40) and in 1900 were able to open a concert hall next door; the Bechstein Hall became a renowned concert venue known for its excellent acoustics. It attracted major performers for recitals, the first of whom on its opening day were pianist Ferruccio Busoni and violinist Eugène Ysaÿe. Sadly, it was also in 1900 that Carl Bechstein passed away. Sons Carl junior and Edwin took over and, by the start of the First World War, it was said that the Bechstein company had the largest piano dealership in Europe.

By this time (with brother Edwin mainly based with the London operation and living there with his wife, Helene), the Bechstein company was patronised by influential tsars of Russia and key royal families across Europe. The company had three factories and turned out 5000 pianos a year. All this was to come to a rather disastrous end, however, once the First World War took a grip. German pianos and music became strongly out of favour (even owning a German breed of dog was unwise; by the end of the war composer Holst had dropped the 'von' from his name); unsurprisingly, royal warrants were withdrawn in 1915, and following the passing of the Trading with the Enemy Amendment Act of 1916, the British arm of the company was wound up. The hall was seized and, after keen bidding, sold at auction to Debenhams for £56,500 (the Paris showroom suffered a similar fate) and the company's 137 pianos were confiscated.

After the war, the firm tried and did recover to a certain extent to regain its former prestige. A promotional film was made in 1926, a new Bechstein Hall was opened in London's West End and Bechstein pianos were used again in some concerts. Well-known pianists loyally endorsed the company; two who took over the reins of the earlier patrons Brahms, Debussy and Liszt, were Alfred Cortot and Artur Schnabel, yet things were never quite the same for the company in Britain (when Schnabel was compelled to use Steinways in America, he wanted technicians to alter the touch and sound to get them to as close as a Bechstein as possible).

Alas, as Hitler rose in popularity and power through the 1920s and 1930s, Edwin Bechstein made no secret of his support for Nazi ideology, perhaps even more so his wife, Helene, a German socialite. She was considered a close friend of the Führer's, even calling him Mein Wölfchen (my Little Wolf) and at an early stage in their relationship was said to be his etiquette tutor. The Bechsteins aided the Nazi party financially and even bought Hitler a new Mercedes Benz car. Edwin died in 1934 and was given a state funeral with close friend Hitler being present. Helene, herself was later sentenced to 60 days hard labour for being a Nazi collaborator. After the Nazi surrender (and their factory was destroyed by Allied bombing, as was the Blüthner piano factory), the company was commandeered by the Allies in the US Occupation Zone and the company was not permitted to start making pianos until 1948. Again, it started the difficult road of rebuilding its name and reputation which, through the excellence and craftsmanship of its pianos, continued to rise. The Bechstein firm today also make W Hoffmann pianos at their second factory in the Czech Republic, while the Bechstein-designed Zimmermann piano is manufactured under licence in China.

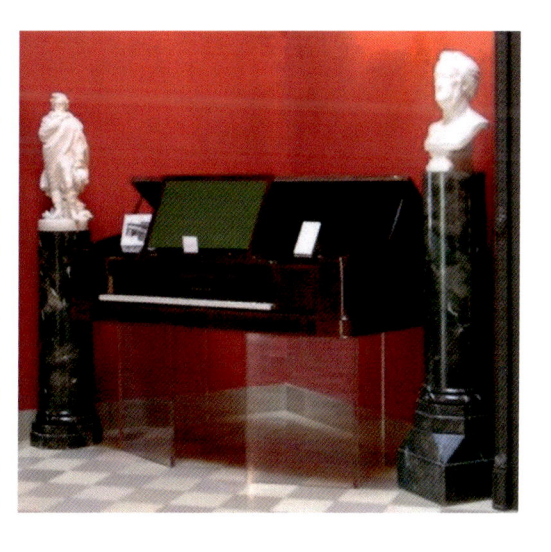

Right: A Bechstein 'composition piano' made for composer Richard Wagner (displayed in his former villa in Bayreuth).

Below: The architect for Bechstein Hall (renamed Wigmore Hall in 1917) was Thomas Edward Collcutt, who was also responsible for the Savoy Hotel. Wigmore Street gets its name from the village and castle in Herefordshire. Currently, a brand-new Bechstein Hall showroom and concert venue is under construction in Wigmore Street.

Bechstein prices of the 1890s: small uprights £60.00, concert grands £300.00.

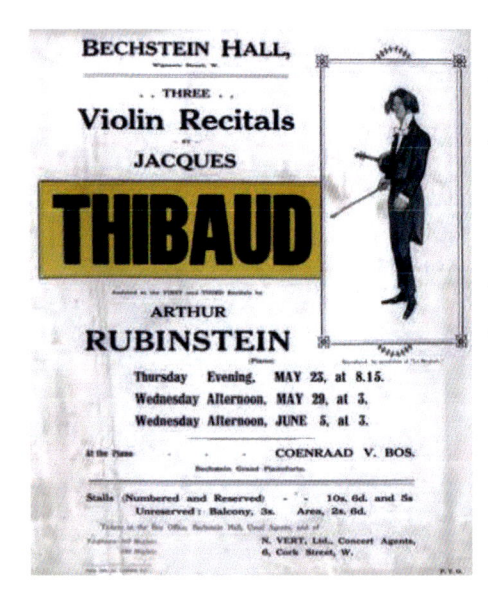

Left: French violinist Jacques Thibaud was a close contemporary of Arthur Rubinstein (though as the billing shows, at that time was more well known). He had been present at Rubinstein's first Bechstein Hall concert and was so impressed went backstage and suggested they do one together. Sadly, Thibaud was killed in an Air France plane crash in 1953 (his 1720 Stradivarius violin was destroyed in the crash).

Although the first London Bechstein Hall was opened in 1900, the company already had a Bechstein Hall in Berlin, where key musical figures such as Brahms and Anton Rubinstein gave concerts.

No connection to the instrument, but there is a fairly rare species of bat known as Bechstein's bat (it can be found in parts of Europe and Asia).

Perhaps the bat mentioned above would be interested in the spider top right. It is, after all, known as the piano-flat spider.

BURNING PIANOS

We know that around the 1960s and later, it wasn't piano hammers but real hammers (and bigger implements) that were being used, yes used in village green piano smashing competitions. Sadly, it was said that pianos were no longer in fashion, no longer needed, which was only half-true, as their popularity did rise again. However, another kind of violence to pianos came in the form – said to be an RAF tradition – of burning pianos. There are different stories that attempt to explain this RAF tradition (which can be seen in the USA and elsewhere also), but it is said that recruits were encouraged to learn to play the piano because it helped their concentration, dexterity and fine motor skills. Added to this is the fact that with so many pilots being killed during the First World War, the RAF was forced to select its pilots from the general population, instead of the preferred upper classes (allegedly more cultured). To avoid such things as piano lessons, apparently, some pilots took to setting certain selected pianos alight (nothing to do with Handel's *Fireworks* suite).

As recently as 2019, the RAF got it wrong when they burnt a piano by mistake. According to Wiltshire newspaper reports, a generous family from Winterbourne Dauntsey, near Salisbury, were downsizing so donated their piano to Boscombe Down military base. They thought it would make a welcome addition to the mess. They later learnt it had been mistakenly destroyed at a Battle of Britain night supervised by firefighters. The station commander apologised for the mistake.

The occasional avant-garde composer, in recent and not so recent years, has used burning pianos in their work. New Zealand composer Annea Lockwood (whose father was an RAF pilot) wrote a piece in 1968. There were some specifics, however, an upright that was beyond repair had to be used. Her piece, *Piano Burning*, entailed the performer dropping a lighter fluid-soaked piece of paper into the piano. Additionally, balloons may be attached, and the piano may be played for as long as the performer is able (*Piano Drowning* is another piece by the same composer, perhaps *Piano Extinguisher* is in the pipeline).

C

COMPOSERS AND THEIR PIANOS; COBBE COLLECTION

When English composer Ralph Vaughan Williams (whose great-uncle was Charles Darwin) was having a lesson in Paris by Ravel, he was told, "You must always have a piano because without a piano, how can you invent new harmonies?" The piano he used when composing (notably *The Lark Ascending*) was a Broadwood upright, which can be seen in his former childhood home at Leith Hill, Surrey – now cared for by the National Trust.

Another English composer, Elgar, perhaps knew more about pianos than other composers and musicians, for Elgar's father and uncle ran a music shop at 10 High Street, Worcester which sold organs and pianos (the family lived above the shop). As a youth, Elgar's father, William, had been apprenticed to London music publisher and piano maker Coventry & Hollier. Sufficiently experienced, as a young man William commenced work as a piano tuner for Stratford's Music Salon in the High Street, Worcester. By 1844, he'd progressed to working freelance, travelling out to his clients wearing a top hat and mounted on his thoroughbred mare. His brother, Henry (who had learnt his piano craft through working with the well-known Kirkman firm in London and then spending time with Hime & Addison's in Manchester), joined him in 1859, when they were able to go into business.

Sir Edward Elgar

Elgar's father, having gone into partnership with his brother, moved into larger premises at 10 High Street. Their prestige perhaps rose after having the dowager Queen Adelaide, widow of William IV, on their books. Both of the brothers were involved in local music, William was a semi-professional violinist and church organist, while Henry played the viola, harmonium and piano. As for Sir Edward, after composing the tune that became *Land of Hope and Glory*, he knew he'd written something special and wrote: *I've got a tune that will knock 'em – knock 'em flat!*

Unlike today, where certain well-known musicians get contracted (or encouraged, coerced, 'bribed' – take your pick) to use one make of piano only, for earlier composers, when pianos were evolving and were less consistent in their performance capabilities, they were more likely to both own and use more than one make of piano. Piano firms, moreover, were frequently keen to present composers and pianists with one of their instruments ('sports' sponsorship when composers and pianists were the sports/rock stars of their time). Franz Liszt had a contract with the French firm Erard, owned a Bechstein, and took possession of Beethoven's Broadwood after his death (though it was too antiquated to use). At some of his concerts, Liszt had two or three pianos on standby, either because each one had tonal qualities more suitable to the piece he was going to play or substitutes were needed due to his powerful playing having the potential for strings to break and actions to fail.

It is often forgotten that in his early career Beethoven was also known as an accomplished pianist, as well as being a composer. Partly because pianos had evolved since the young Mozart's time, Beethoven's piano technique was in contrast to Mozart's (though Beethoven was always in awe of the great composer). A music critic of the time, Mosel commented:

A year after the appearance of 'The Magic Flute,' a star of the first magnitude rose above Vienna's musical horizon. Beethoven came hither, and attracted general attention as a pianist even then. We had already lost Mozart; all the more welcome, therefore, was a new and so admirable artist on the same instrument. True, an important difference was apparent in the style of these two; the roundness, tranquillity, and delicacy of Mozart's style were foreign to the new virtuoso; on the other hand, his enhanced vigour and fiery expression affected every listener.

The Broadwood square piano Edward Elgar acquired as a reconditioned instrument from his father's stock in the 1890s (it was one of Broadwood's school models). He used the six-octave piano at his Malvern cottage. On the soundboard can be seen the titles inscribed by Elgar of compositions he had just completed. It is known that the piano was used from start to finish while Elgar composed the *Enigma Variations*.

Rock musician Sir Brian May seated at the Broadwood piano once owned by Gustav Holst (and now in the Holst Museum). May is a fan of *The Planets* suite, in particular *Saturn*. Holst wrote a preliminary two piano version of the work. Early reviews of the suite were not all positive, the *Sunday Times* describing it as 'pompous and noisy.'

Cole Porter's 1907 Steinway model B grand was gifted to him by the Waldorf Astoria hotel in New York where, in later life he had an apartment on the 33rd floor for around thirty years. It has recently been fully restored by Steinway, but to Cole Porter he lovingly called it his *High Society* piano, for he wrote *Anything Goes*, *Night and Day*

and countless other classic numbers on it (including the lyrics). Some sources tell us: when he attended his first main school, he was allowed to take a piano with

him; while at Harvard Law School, with the agreement of the dean, he transferred to Harvard's music department. It is also reputed that Porter served in the French Foreign Legion and that he carried a specially constructed portable piano around with him so that he could entertain the troops in their bivouacs. Lastly, in 1937 he suffered a serious riding accident when his horse fell on him, crushing both his legs. Some years later he had to have his right leg amputated. His Steinway was mounted on special blocks so that he could access the keyboard while in his wheelchair. When completed in 1931, the Waldorf was the tallest hotel in the world; it still has a Cole Porter suite.

Left: *c.*1905, composer and pianist Sergei Rachmaninoff is seated at his Blüthner grand. According to one of his piano tuners, the late William ('Bill') Hupfer, the composer was a lover of speedboats and cars – he was the first in his neighbourhood to own a car. Before concerts, he would ask Hupfer to drive him there. By the end of the journey, the careful tuner-driver had been pushed out and replaced by the rather impatient and speed-loving composer! Rachmaninoff was also known for his large hands and could stretch thirteen notes.

Below: Two of Liszt's pianos in the Liszt Museum, Budapest.

THE COBBE COLLECTION

The Cobbe Collection is housed in the National Trust owned country house at Hatchlands Park, situated near Guildford in Surrey. The musical instrument collection contains around forty-two early keyboards, many having once been owned by famous composers. The collection includes harpsichords, fortepianos and square pianos (the latter are actually rectangular in shape; early square pianos from England proved to be highly fashionable in Paris). There are three of Frédéric Chopin's pianos in the collection: an Erard, Broadwood and Pleyel grand built in 1848. He had these three pianos in his drawing room at 48 Dover Street, when a visitor to London and described his temporary Mayfair lodgings as 'piano heaven'. From time to time, concerts are given in the main music room in Hatchlands Park and utilize many of the instruments. The pianos on display include:

JC Bach's square piano: Johannes Zumpe & Gabriel Buntebart (1777- 1778)

Mahler's grand piano: Conrad Graf (*c.* 1836)

Elgar's mahogany square piano: John Broadwood and Sons (1844)

Bizet's composing table piano: Roller (1855)

Marie Antoinette's square piano: Erard (1786-1787, shown below)

Below: To begin with, square pianos often had no pedals though some may have had stops or levers to control the dampers.

As the instrument evolved, some were built with just one single (sustaining) pedal but it was designed to be used by the left foot. Some later squares had up to five pedals, each used to create different effects.

Chopin's treasured Pleyel grand. The composer was a friend of Camille Pleyel and a great fan of his pianos. Chopin paid 2200 francs for the piano; when collector Alec Cobbe acquired it from a dealer in 1987, he paid £2000. They had no idea then that it was Chopin's piano.

Above and right: Italian upright loaned to Liszt by maker Carlo Ducci when he visited Florence. It is thought to be the only Ducci piano still in existence.

Left: Portrait of Joseph Haydn by the English artist Thomas Hardy (the painting is in the care of the Royal College of Music).

Left: Haydn's fortepiano of 1794-95. Made by London firm Longman and Broderip, it has a compass range of 5½ octaves (he took it with him when he returned to Vienna).

D

DAMPERS; DICKENS; *DESERT ISLAND DISCS*; DIBDIN; DIGITAL

The individual dampers on a piano increase in size as the strings get longer from the treble down to the bass. Each wooden damper contains a pad of felt that rests on the strings and stops them vibrating until a note is played (as the hammer travels towards the strings, the damper is lifted off the strings before they are struck; when the key is released, the damper returns to its resting position on the strings). On grands the dampers rest on top of the strings, in a modern upright they have small springs which push the dampers on to the strings. When the sustaining pedal is pressed down, all the dampers are lifted off the strings at the same time. For about the last twenty notes (depending on the piano) into the top treble section, pianos do not have any dampers mainly because the strings are shorter and do not vibrate very loudly or for very long. In a small minority of pianos, the treble notes which do not have dampers may tend to vibrate after a note has been played. As it might be a minor irritation to some pianists, it is worth checking how far the dampers are in place in the treble section and trying the piano out to confirm that all notes are damped promptly after a note has been played and that no after-sound can be heard. Note: the white-felted dampers can be clearly seen in the picture of a prepared grand on page 136 (notice how they don't continue right up into the very top treble section).

DICKENS

Charles Dickens' favourite composers were Mendelssohn, Chopin and Mozart, and although he wrote the occasional song and enjoyed music (also owning two pianos in his lifetime), Dickens wasn't known to be a competent musician. His sister Fanny, on the other hand, won a scholarship to the Royal Academy of Music (she was sponsored by royal piano maker Thomas Tomkison). A letter to the *Daily News* from an old school friend tells us that Dickens had tried to learn music, but that one day their music master had given up teaching him the piano, declaring, 'He had no aptitude for music, and it was robbing his parents to continue giving him lessons.' That said, it was known Dickens could turn his hand to a bit of accordion playing.

DESERT ISLAND DISCS

Radio 4's programme, *Desert Island Discs*, has been running almost continuously since 1942. The first guest was comedian and musician Vic Oliver, whose first chosen piece of music was Chopin's Étude Op. 10, No. 12

in C minor, played by Alfred Cortot. The programme's format changed over the years and, by the time Richard Ingrams (one-time editor of *Private Eye*) was invited on as a guest, he was allowed to take not only the Bible and the complete works of Shakespeare, but a book of his own choice. For the one luxury item all 'castaways' are allowed to choose, he opted for a grand piano. With foresight, the book he chose was *Teach Yourself Piano Tuning*. (It appears he forgot to ask for a small supply of piano tuning tools.)

By 2019, for the castaways the most common luxury item chosen was paper or writing materials. A piano was the second most common luxury item requested, followed by a guitar. Not all pianists chose a piano, Arthur Rubinstein chose a revolver; Victoria Wood said she was obsessed with the piano as a child, but chose a bumper book of Sudoku, with blank pages and pens. Concerning pianos, Daniel Barenboim wanted a piano with a mattress; Sir John Mills wanted *his* piano; Sir David Attenborough wanted a large, very expensive grand piano; James Stewart, Julie Andrews, Phil Collins, Jools Holland and Sir Anthony Hopkins merely asked for 'a piano'; Dame Fanny Waterman wanted a grand piano and stool; and John Lill wanted a solar-powered piano. Additionally, sixty-one castaways asked for a bed, while golf clubs, radio receivers and various forms of alcohol were requested by lots of male castaways. Lastly, for pieces of music, Beethoven had the most favourites to his name but the most chosen piece was Handel's *Messiah*.

DIBDIN (1745 – 1814)

Baptised in Southampton's Holyrood Church (see tablet, left), musician and prolific songwriter Charles Dibdin, having been a Winchester chorister, first came to London in his young teens and was employed to tune harpsichords in a Cheapside warehouse. His *Tom Bowling* is sometimes performed at the Last Night of the Proms, but he is also credited as the first person to publicly demonstrate the piano in London (if not the country), which he did between the acts of a performance of *The Beggar's Opera* at Covent Garden on 16 May 1767, where he accompanied soprano Miss Brickler on the 'new piano forte'. (The following year, JS Bach's son, Johann Christian, having purchased a square piano from Zumpe, performed a piano solo at a concert in London on 2 June.) Dibdin lived in Camden Town's Arlington Street, an area which later became synonymous with English piano manufacture. The politician Baron Heseltine (Michael Ray Dibdin) is a fan and distant relative. The popularity of the piano saw published in London in 1797, *The Pianoforte* magazine.

DIGITAL PIANOS

Electric and digital pianos, what predictions can we make for a hundred years into the future? Will the modern, powerful acoustic concert grand only be seen in museums or heard very occasionally by small audiences being played by early instrument buffs, as fortepianos are today? The strong and heavy cast iron frame that was eventually seen in all modern pianos, was at first greeted with doubt and suspicion. How can this heavy chunk of tackle be seen as musical, it doesn't add anything to the instrument in any way, does it? Wood, often very decorative wood, had been used by craftsmen for centuries in harpsichords and then the fortepiano. Unwieldy iron was hardly likely to catch on. Yet it did and, along with improvements to the piano's action, revolutionized the piano and the music composed for it. Slow and steady, however, the electric piano made its entrance from the 1920s, followed by the digital piano in this century. When people talk of the piano a hundred years from now, will they automatically think of the digital piano rather than the somewhat forgotten primitive acoustic piano that people, it was said, used to own and even enjoyed playing? History will reveal all.

A YouTube Pathé News clip of a Neo-Bechstein electric grand being played in the early 1930s is fascinating to watch. The piano had all the conventional piano parts but required no soundboard as electrical pickups were used instead. The Czech firm, Petrof, later acquired a licence to manufacture their own version.

The Whaletone grand is at the cutting edge of digital piano design. It can be played as a normal piano (though does not require the standard piano parts; it also has digital play-back facility). These new and innovative digital pianos take around 6 to 9 months to make. Sales are world-wide but the company is based in Europe.

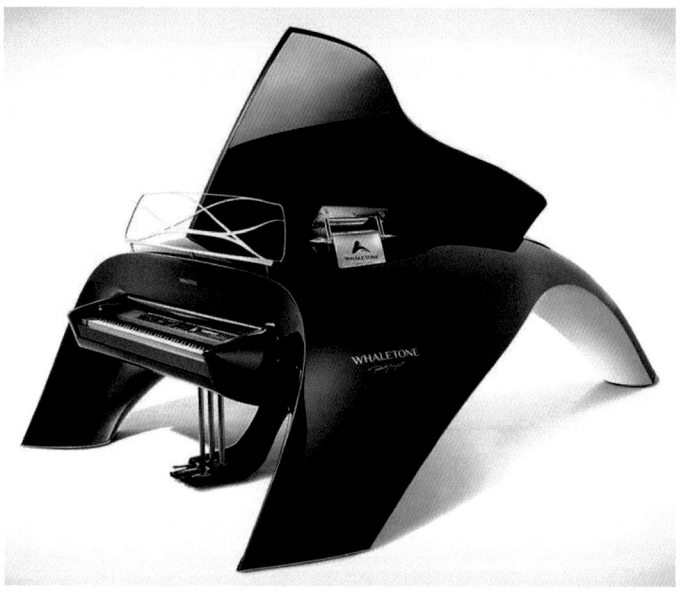

E

EIGHTY-EIGHT; EQUAL TEMPERAMENT

Eighty-eight keys, that's how many keys contemporary uprights and grands have though, often, several decades ago the standard for most pianos was a range that consisted of A in the bass to top A in the treble: eighty-five notes/seven octaves. Grieg's famous piano concerto in A minor (just the one, a second concerto was never completed) has an opening that first incorporates the top A of the piano and soon after uses the lowest A note in the bass. It's an exciting opening, as pianist Stephen Kovacevich tells us:

Mentally speaking, the opening is the hardest part: walking on stage, sitting down, hearing the timpani roll and knowing you have to deliver the goods. I've heard the most accurate pianist of our time make a small mistake in performance – and every pianist in the hall sighed with relief. (But I don't think the audience even noticed.)

Above: Liberace's car, complete with special number plate (pianist Paderewski became a family friend). Before gaining fame, he performed under the name Walter Busterkeys. Overleaf can be seen his eighty-eight notes piano-shaped swimming pool (though it is said Frank Sinatra had one first).

Once a year around the world, Piano Day is celebrated on the 88th day in the year. There are today some makers who have built grand pianos with a wider compass, the Bösendorfer 209 has 97 notes (see also page 186). The Erard company, similarly, made pianos with 90 notes as early as 1877.

His signature included a piano; his watch was also shaped as a piano. To tell the time he had to lift a miniature grand piano lid.

This piano-shaped pool is part of the Walt Disney organization in Florida.

Reputedly, it was the Steinway company who first started making pianos with 88 keys, which they began doing in the 1880s. Oddly, after the second world war the Broadwood company had to make an upright with only 87 keys for Cunard's refitted *Queen Mary* because of space limitations in getting it aboard. Currently, Australian firm Stuart & Sons in New South Wales are genuine piano builders who include in their range grand pianos with 102 keys – they have also built a record-breaking 108-key/nine-octave grand.

EQUAL TEMPERAMENT

In many parts of the world, the music nearly all of us hear around us in our everyday lives (recorded or live) uses a tuning system that has been a default system used for around 200 years. It has a name but is not something most people know about, the tuning system is called equal temperament. Books have been written about it, there is plenty of science and maths to go with it (connected to the Pythagorean comma and circle of fifths), so what follows is more of a shorter layman's explanation.

For musicians who can tune their own instruments, such as a violinist, they can play a musical interval (for example, the major third C to E) how they want to. Being musicians, they would naturally play this interval musically correct, usually as a pure ('just') interval. Every interval on the piano has to be tuned using compromises; apart from octaves, every interval in one chromatic scale in the middle of the piano has intervals that have been deliberately tuned so they are not perfect/pure but the gap (or frequency rate) between each note is consistent and evenly spaced (and the rest of the piano is tuned in octaves from this first tuned/readjusted chromatic scale). On the piano, the major third C to E just mentioned, is tuned wide (sharp), not that most people would notice because the equal temperament tuning system has been used and heard for many generations and has become the norm. So there is an oxymoron here, pianos tuned to equal temperament are both in tune and (theoretically) out of tune as none of the intervals, barring the octaves, are perfect/pure (even the octaves are not always put perfectly in tune, they are stretched to enhance the sound).

So why, on the piano and other fixed pitch instruments such as organs and xylophones, are the musical intervals not tuned to be pure intervals? The keyboard is a compromise with the black notes doubling up to serve as two notes. For example, a C sharp on the keyboard is physically also a D flat. With singers (not singing along to a keyboard instrument) and other instrumentalists who can control the tuning of their instruments, musically C sharp and D flat are not quite the same note; depending on the piece being played, there would be a difference between these two notes (and with the others; for example, G sharp wouldn't be the same note as A flat as it is on the piano). To have actual keys/notes for all the sharp and flat notes on a keyboard would make it rather unmanageable to play. If, on the other hand, that interval C to E *was* tuned perfectly as a pure sound, it would be at the expense of the other intervals in that middle chromatic scale having to be adjusted more severely and the octaves not working out. In other words, all intervals in the one octave on the piano are tuned – that is adjusted – so that they sound musical and are evenly spaced, and although none of the intervals are pure, they all 'get on with each other', with none sounding wildly out of tune (in effect, the tuner is trying to sit thirteen people fairly on twelve seats). If the tuner started to tune the first major third, middle C to

E, purely, tuned the next major third, E to G sharp purely, then finished the octave by tuning the following major third G sharp to the C above middle C, on a piano the octave wouldn't work out, the higher C wouldn't match with the lower (middle) C, thus these intervals and all the others have to be stretched or narrowed to make a usable chromatic scale and octave.

For equal temperament, the fifths are tuned almost pure, just slightly narrow/flat. The fourths are slightly wide/sharp, but the major thirds are the most noticeably sharp. That said, it is the standard tuning system our ears have been conditioned to. Because a piano has notes which contain three strings throughout much of the piano, the strings on each note must be perfectly in tune, tuned as unisons. A piano has in the region of 230 strings, if it has been tuned correctly but sounds as though it is beginning to go out of tune, it's not so much its lack of equal temperament tuning one would hear, the out-of-tune sound is likely to be from the unisons and octaves losing their pitch and being out of tune (sounding rather 'honky-tonk').

Singers and other instrumentalists can do more with tuning and hitting notes, they can bend notes, use vibrato and slide up or down a pitch. String players, when not playing with a piano, can play more musically pure (to some ears it might sound more pleasing, to others, the less dissonance might make it sound rather bland, it's a matter of taste, not necessarily because just/pure intonation is always better). Instruments playing with a keyboard – whole orchestras when it's a piano concerto – first tune their instruments to the piano and then, by default, play using equal temperament so as to be in tune with the piano.

Much of the Western world began to use – or attempted to use – equal temperament (or a system close to it) not long after the first pianos came on to the market because it was a convenient system. Prior to equal temperament gaining widespread use, other temperaments were in existence. Harpsichords and organs were often tuned using the meantone temperament system where more of the intervals were tuned purer (when no beats – dissonance – can be heard, but with this temperament not every interval can be tuned completely purely). With the meantone and other temperaments, playing music in related keys might be okay and give each key its own character, but straying into more distant keys would show up too many clashes where certain intervals had a noticeable 'wolf': the beats being too fast, making the interval dissonant and sounding out of tune. (With the meantone temperament, if one compared playing C with its neighbour C# using equal temperament and meantone temperament, for the latter the interval would sound slightly narrower/flatter.) It is often erroneously stated, incidentally, that JS Bach wrote his *The Well-Tempered Clavier* (which uses all 24 keys) to show off equal temperament, but this is not the case. Equal temperament was already known about before Bach's time though simply wasn't in every-day use when he was alive. It is said

Bach used something close to it, for example the temperament known as Werckmeister, but it is known he tuned his own harpsichords and it is likely he experimented with various temperaments.

It is easy to forget that there are musical notes 'between the notes', it's just that Western music doesn't use them. On a keyboard, the smallest gap between one note and the next, is a semitone (a half note). In other musical genres there are such things as microtones – the notes that are there 'between the cracks'. One exponent of this musical system is the late Lou Harrison, an American composer and one-time student of Arnold Schoenberg (who had a great fear of the number 13; he was born on the 13th and, despite purposely staying in bed all day, died on the 13th – a Friday!). A great fan of the microtone, Harrison liked to say, "Equal temperament destroys everything and is not for the human ear."

Similarly, there are enthusiasts of early keyboard instruments who prefer to tune their instruments using one of the earlier pitches and temperament tuning systems (meantone for example). Part of their argument is that composers wrote their music for instruments which, at the time, used a non-equal temperament tuning system and when music is played on more authentic instruments tuned to an earlier kind of pitch and temperament, you are much closer to hearing what the composer heard. It is argued that musical keys have their own colour and characteristics, much of which is lost when instruments are tuned to equal temperament. Other temperaments can be used for tuning an instrument specifically for a certain piece of music. To some, this tuning system might well enhance the characteristics of the piece, allowing the listener to hear it in a different light, but the tuning system might let itself down should the performer venture into other keys. That said, it may well be the case that should an instrument be tuned using a non-equal temperament system, the average listener might not notice that it wasn't tuned using equal temperament (others might detect that it sounds different but not necessarily that it is 'bad', better or, even, out of tune). For everyday use however, equal temperament is more versatile and allows for the greatest range of music, in any key, to be performed without having unsavoury beats between certain intervals.

If we were to compare the figures and ratios between intervals tuned to equal temperament and those tuned to a just scale, it would be noticeable that some intervals are tuned quite sharp (as with the major third) while others are less affected and tuned close to perfect intervals. The fifth is the least affected interval by equal temperament, but how much to sharpen or flatten any interval when piano tuners lay their middle scale using equal temperament is the fundamental skill that can take them several years before good accuracy can be achieved. The fifth is only very slightly flattened as any deviation from a pure interval is very noticeable (emitting a wolf or 'wow-wow' set of beats which many people would find unmusical).

F

FRAME; FINCHCOCKS; FORTEPIANO; FAMOUS PIANISTS (CLASSICAL); FAMOUS PIANOS; FURNITURE

Called the harp or plate in America, all modern acoustic pianos have a one-piece cast iron frame (the modern iron-framed piano evolved from wooden-framed early pianos to those then having iron reinforcements and plating to take the strain of increasing string tensions). The iron frame is what gives the piano its considerable weight, a baby grand without its frame could be easily lifted and moved by most adults. In a grand piano, the frame rests on a shelf inside the piano's rim and on top of the soundboard; it helps considerably to keep the tuning of the instrument stable (and holds around twenty tons of string tension). Better quality pianos always have the piano manufacturer's name (or emblem) on the frame, whereas older and cheaper mass-produced pianos tend not to have any visible name on the frame.

The gold frame of a grand piano (the iron frame weighs around 450 lbs). Larger concert grands can weigh around 1,200 lbs (544kg) – in the same region as the weight of a large horse. It was American maker Alpheus Babcock who first patented a cast iron frame in 1825.

FINCHCOCKS

Finchcocks in Goudhurst, Kent, is a country house that was built on the site of an earlier property and was completed in 1725. Up to 1796, the property had been in the Bathurst family for 230 years but was then sold to the Springett family. Poet Siegfried Sassoon was a regular visitor in the early 1900s. In the Second Word War it was used by pupils and staff of the long-established King's School in Rochester (the oldest choir school in the world) until it was later requisitioned by the army. By the 1960s it was home to the Legat Ballet School. Later on, from the 1970s, Finchcocks was purchased by the concert pianist and early keyboard collector, the late Richard Burnett MBE, and run as a musical museum and charity. Popular period concerts were also put on (sometimes with actress Prunella Scales in role as Victoria). Part of the keyboard collection and Finchcocks charity is still run by Richard Burnett's wife, Katrina Hendrey. Currently, Finchcocks is also used for research, concerts and numerous courses (some residential) on piano teaching and performance – each course member gets a grand piano to use.

With Georgian architecture, the house today is a Grade I listed building; a modern view can be seen overleaf.

Right: A 'lyre' upright piano made in Berlin, *c.*1830, part of the Finchcocks collection.

A practice grand used by the Finchcocks piano school.

FORTEPIANO

Fortepiano, or should that be forte piano (possibly hyphenated?), piano forte, pianoforte or simply piano (or Joanna for those born within the sounds of Bow Bells)? 'Piano' is the commonly used term today, and fortepianos are of course pianos too, but the term has been used by different people at different times, sometimes being dropped or reversed by the same people as the piano evolved.

The name 'fortepiano' was and is often used loosely for some of the earliest pianos that were built. In more recent times, the term has been used by early music enthusiasts and those wanting a more technically accurate reference for the first pianos which evolved from the harpsichord, they had small leather-covered hammers to strike the strings. These first pianos followed after Italian inventor Bartolomeo Cristofori's first pianos of *c.*1709. The interior of these fortepianos had a wooden frame to which the strings were attached, though the pianos came in different guises as regards compass range and pedals or levers. Whilst there were, later on, other kinds of piano also being made, for example the square and, later still, upright, it is generally accepted that the fortepiano – a small grand-type piano with a wooden frame to support the strings – existed until around *c*.1830s.

In practice, the term fortepiano can cause confusion as, at the time and later, it tended to be used interchangeably (so also 'piano forte') for anything the user perceived to be a piano. The earliest pianos built by Cristofori and the other early makers who followed in his footsteps (one being the great harpsichord and organ builder Gottfried Silbermann – having completed his grand organ in Freiberg Cathedral, he diversified and began making fortepianos for Frederick the Great) were, of course, 'pianos' because they all had strings which were struck by hammers rather than being plucked by individual plectrums (as in the long-established harpsichord). But both the upright and grand piano we have had for the past two centuries or so aren't normally referred to as fortepianos because they are rather more different in design, construction and sound.

Horizontal and with a winged-lid, musically the fortepiano was the relatively small and lightweight instrument upon which composers such as Haydn and Mozart composed their music (though some fortepianos did later on resemble the fuller grand piano). It is from this time that we begin to see music written specifically for the piano, and this music and the concerts where pianists started to play this music helped to make the instrument popular (thus making greater demands on the instrument and prompting its evolution from a wooden framed instrument to a robust iron framed one with a larger keyboard range and more versatile action).

When the first pianos were built, 'fortepiano' also had currency in France and Germany. A manuscript of Beethoven's auctioned in 1913, moreover, had written in his own hand: Quintett fürs Fortepiano. Other composers wrote compositions specifically for the fortepiano, and, in the 1790s, some musical instrument stores were advertising both harpsichords and fortepianos, but elsewhere the term pianoforte was gaining currency. An 1825 programme for a benefit concert held at Bristol's Theatre Royal announced that, at considerable expense: *the celebrated Piano-Fortest, Mr Moscheles, has been engaged to play the piano-forte* (adding: *just arrived from the Continent*). Pianist and composer Ignaz Moscheles became a teacher and close friend of Mendelssohn. By the beginning of the Victorian period, pianoforte had gained common usage. Even before this, references to the pianoforte can be seen in the works of Charles Dickens and Jane Austen (in Austen's novel, *Emma*, she adds authenticity by mentioning the name of the grand piano, a Broadwood; Austen herself played her own piano daily before breakfast).

It is more from the middle of the last century that the word fortepiano was revived and applied to either the original or reproduction pianos from the time when such composers as Mozart and Haydn were around. This revival of the fortepiano coincided with enthusiasts either wanting to hear or perform on the kind of instruments the earliest composers of piano music would have been familiar with.

'Period' concerts from the 1950s, sometimes by well-known artistes at important venues, were not uncommon and continue today (not forgetting recordings also).

So how was the fortepiano different to later pianofortes? The earliest fortepianos had a smaller and lighter case/body. They had smaller, leather-covered hammers than later pianos (which used compressed felt), therefore giving them a lighter sound and touch (the touch/key dip was also shallower than on later pianos). Whereas the pianos that followed had trichords (three strings to each note throughout much of the instrument), some early fortepianos had only bichords. It is understandable, then, that the fortepiano would have had a different feel and tone to later pianos, especially when one remembers they did not have the important structural feature of an iron frame. The move away from wooden frames saw the addition of metal plates which evolved into the one-piece cast-iron frame still used today. This helped to prevent pianos warping and to comfortably accommodate more strings, also allowing them to be tuned at a higher tension; importantly, it gave the piano's tuning more stability over a longer period of time because the iron frame was less susceptible to changes in temperature and humidity. Another significant improvement on the earliest pianos, in addition to the keyboard's compass increasing in range, was an improvement in the design and capabilities of the piano's action. The principles of Cristofori's first escapement action were refined yet some of the basic features of Cristofori's piano actions can still be recognized in modern grand pianos today. We draw this piece to a close with a concert review by Joan Chissell, published in *The Times* newspaper on 29 November 1967:

Telephones may be better than pigeon-carriers, electric light better than oil lamps. But in the sphere of the keyboard, progress is more debatable. Or so it seemed last night in the Purcell Room. When Joan Davies drew such lovely sonority from a fortepiano (built by Arnold Dolmetsch) that the modern pianoforte she used after the interval sounded almost retrogressive, contributing little extra to human happiness but volume. Beethoven is the Titan first thought to need the bigger monster...

There are enthusiasts and orchestras today who perform music on original or replica instruments. One, The Orchestra of the Enlightenment (which has an early keyboard section), performs at major concert venues.

Left: Made in Vienna, this is a Conrad Graf on display at the Metropolitan Museum of Art. Built in 1838, it is a later fortepiano (quite solid looking though doesn't have an iron frame). About a year after this piano was made, composer Robert Schumann acquired a similar model from the same firm. Chopin, too, visited the Graf piano shop daily to practise on one when he was staying in Vienna and performing in concerts.

FAMOUS PIANISTS (CLASSICAL – A FEW OF THE MANY)

Perhaps 'important' would be a better term, for what shortly follows is a biographical sketch of just a few of the many classical pianists who became household names (not here presented in order of greatness, nor are they all necessarily world famous, yet they may still be interesting and significant).

Certain great pianists were often tied to particular makes of piano, so had to be careful what they said about the make they were officially or unofficially contracted to perform on and also what they thought about other makes. Moreover, buyers of pianos often forget that, occasionally, a second-hand piano from a top manufacturer might not be a particularly pleasant instrument at all. Similarly, other less well-known makes have often turned out to be wonderful instruments in both tone and performance. Each instrument needs to be judged on its own merits as no two pianos are ever the same even if they are the same model from the same manufacturer and built in the same year. Arthur Rubinstein was closely associated with Steinways, but earlier in his career had performed on Bechsteins and other brands. Indeed, in the second of his autobiographies (*My Many Years*), he wrote about a Blüthner piano he came across and used for a recording:

After my coffee and cigar we went to one of the recording rooms where they had a Blüthner piano. Well this Blüthner had the most beautiful singing tone I had ever found. I became quite enthusiastic and decided to play my beloved Barcarolle of Chopin. The piano inspired me. I don't think I ever played better in my life.

Unofficially, Rachmaninoff was something of a Blüthner fan (it seems to be an Americanism to spell his name with a 'v' at the end, though his grave in America uses the double 'ff' spelling). He wrote, 'There are only two things which I took with me on my way to America. My wife and my precious Blüthner.'

The pianist Horowitz once said, "There are three types of pianist: Jewish, gay and bad." What follows below are a few biographical sketches of the many famous classical pianists who have entertained audiences in Britain and, often, globally.

Left: Statue of Polish pianist Paderewski in Ciężkowice, Poland. In addition to other statues of him around the world, he would have schools, festivals and even vodka named after him.

Ignacy Jan Paderewski (1860 – 1941)

The legendary Polish pianist Paderewski was said to have amazing magnetism and stage presence. Even now he seems something of a household name even though very few people would have actually seen or heard any of his performances. It is easy to forget, too, that the earlier concert pianists were A-list celebrities. When there weren't the famous film stars and sports people seen in later decades, and not the availability of good quality recordings, people clamoured to both see and hear these early 'rock stars' at live performances. It must have been odd, too, for someone such as Paderewski not to be able to actually hear again what he had just played, for early in his career he also would not have had the opportunity to listen to his concert performances on good recording equipment (if at all).

Born in 1860, Paderewski's mother died soon after he was born (his own wife, sadly, would also die very early on, soon after giving birth to their severely handicapped son). Paderewski was mostly brought up by relatives, but after having piano lessons early on, eventually found a place at the Warsaw Conservatory.

He married his second wife, a Polish actress, in 1884. Although still studying composition, and also teaching, his career as a concert pianist took off and he toured extensively. In fact, he would tour America over thirty times and eventually made it his home (he would be the first pianist to give a solo performance at the new Carnegie Hall). 'Paddy mania' had long since been established for the pianist after just a few visits there, yet the hysteria surrounding Paderewski is said to have started in England, at a concert in St. James's Hall, London in 1890. Possibly he was the first concert pianist (and only?) to have female fans practically throwing their knickers at him, as historian Ludwig Stomma reveals:

When Paderewski appeared in the entrance, the crowd was taken over by hysteria. Young townswomen of London, whom normally one would not have suspected even a shade of temperament, were now pushing towards the staircase, trampling each other, and shouting: 'Take us! We belong to you! We are yours!' Next, they pulled up their skirts to reveal frivolous underwear with the name of their idol embroidered next to a red heart. For the sake of public morals, the police had to intervene.

Although he became wealthy, Paderewski also became a philanthropist and helped many causes. For his own compatriots, he campaigned tirelessly on political issues; indeed, he would be elected Poland's prime minister in 1919. Although his tenure was very short, not quite a year, his government achieved a great deal, and his influential political involvement with his homeland had started before 1919 and would continue after he left office. Among the significant milestones his government achieved in such a short time, are included: democratic elections to Parliament, ratification of the Treaty of Versailles, passage of the treaty on protection of ethnic minorities in the new state, and the establishment of a public educational system.

Paderewski is unusual in being not only a prime minister who was a concert pianist, but one who wrote, among his other compositions, both an opera and a piano concerto. Admired for his prodigious pianistic skills, there were those who saw him more as a showman. To his compatriot, Arthur Rubinstein, he was a marvellous personality and speaker rather than a marvellous concert pianist. When Paderewski toured America during the 1920s as a world-famous pianist, incidentally, he travelled in his own railway car, complete with cook, piano and piano tuner.

Having once been the most sought-after pianist in the world, sadly his life came to an end after suffering from pneumonia during the early part of the Second World War. He was buried in Arlington National Cemetery, though his remains were returned to Poland in 1992 and interred in St. John's Cathedral, Warsaw. He may have gone, but he would never be forgotten, as awards continued to be bestowed on him, museums were named in his honour, a USA 1960 postage stamp carried his still very familiar features with flowing red hair, and he was given his own star on Hollywood's Walk of Fame (by his seventies he had become a movie star too).

Arthur Rubinstein (1888 – 1982)

Perhaps we shouldn't be surprised that the Polish-American pianist, Rubinstein, would be a child prodigy at the age of four. It was said that he already had perfect pitch when aged two, and although his father, who owned a small textile factory, had encouraged him to take up the violin, his young son showed an almost obsessive interest in the piano after watching his sister take lessons (he pointedly smashed to pieces a violin he was given).

Born in Łódź, Poland, in 1888, Arthur (Artur in his native language) Rubinstein grew up to be multi-talented. His acute ear for music, also a remarkable memory, helped him to master around six different languages. While walking and breathing music on a daily basis – he saw the music he carried around in his head as akin to having an extra limb – he did have wider interests also, particularly in the arts. For example, at the age of forty-five he married Polish ballerina, Nela Mlynarski; and when living partly in Paris, he became friends with Picasso, but he also had a love of films and would appear in some of them too.

The youngest of seven children, early family life appears to have been happy enough (despite his father going bankrupt). Rubinstein would grow up to be immensely proud of his Polish and Jewish roots, though claimed to be an agnostic.

When aged ten, Rubinstein moved to Berlin to continue his education. It was here that he gave his first performance with the Berlin Philharmonic in 1900, aged only thirteen. Germany, however, in the ensuing years, proved to be a depressing time for the pianist. With a love interest that worked out unfavourably and mounting debts – at the lowest point he couldn't even pay his hotel bills – he attempted suicide. Yet he quickly saw his slightly bungled attempt as a positive and from then on lived up to a motto he'd earlier adopted: Never give in, be courageous in life.

There would be later performances in Germany, but not after the Second World War. Having lost family members in the holocaust, he vowed never to perform there again, telling interviewers "There are two countries where I haven't played. Tibet because it is too high, and Germany because it is too low."

He would become an acclaimed expert in the interpretation and performance of Chopin's piano repertoire, but of course also played most of the other well-known piano works – including concerti – by other composers. It is easy to forget, too, that he was a masterful pianist for chamber work and often performed with leading string players. His early musical training, however, had been under the pianist Karl Heinrich Barth. And what an impressive musical lineage that was: Barth had been taught by pianists who had been taught by Liszt, and Liszt had been taught by Czerny, a one-time pupil of Beethoven. Barth would get to know the composer Brahms personally, yet Rubinstein was born in an era that saw him meeting and, on occasions, socializing with numerous composers, including; Ravel, Poulenc, Prokofiev, Stravinsky and Rachmaninoff.

1904 would see Rubinstein move to Paris, where he launched his career. It was here that he met Saint-Saëns and actually played the second piano concerto in the composer's presence (it would prove to be Rubinstein's warhorse). He wasn't to know it then, but he was destined to divide his family living arrangements, having homes in both Paris and America (he became a naturalized American citizen in 1946).

Rubinstein arrived in London in 1912, where he made a successful Bechstein Hall (later renamed Wigmore Hall) debut. He was fortunate in being welcomed at the Chelsea home of Paul and Muriel Draper. Paul Draper was an American lieder singer, and his wife Muriel was active both in the arts and politically. Their home became a meeting place for key figures of the time, including Stravinsky, Osbert Sitwell, John Singer Sargent and Pablo Casals.

In his long career, Rubinstein toured extensively, but early on and while living in England during the First World War, he began tours in Spain and South America (unlike certain other artistes, he mostly enjoyed travelling, hotels and seeing new parts of the world). No doubt he enjoyed, too, the great welcome the enthusiastic audiences gave the 'new' maestro on these early tours. Such a contrast to his pre-war experiences and that suicide attempt. He'd made his debut in America at the Carnegie Hall in 1906, and the tours that followed it had not been particularly successful, ultimately leading to climbing debts and depression.

A rather closer to home stellar figure, almost rival, was the younger Russian pianist Vladimir Horowitz (who would also become a naturalized American citizen, as did both Godowsky and Rachmaninoff). Early on, the two became friendly and even sometimes socialized together. Rubinstein at that time saw Horowitz as being 'all the rage' and admitted to having a pang of jealousy over his popularity. The two were once invited by Alexander Steinert to dinner, where they enjoyed each other's company. There were two pianos available, both of which were put to good use by the young pianists playing a wide range of repertoire and pieces for two pianos.

At some point in their early relationship, a minor tiff or opinion about a comment grew into an on-off dislike of each other (Horowitz's piano tuner was never allowed to mention the name Rubinstein in the maestro's house). In Rubinstein's eyes, he felt Horowitz viewed their friendship as 'a king for his subject' and although they had become friends and consulted each other from time to time over encore suggestions and other musical matters, it was not a friendship on equal terms. It is evident from Rubinstein's memoirs that, for a time, he really did feel inferior to Horowitz (though ultimately he was inspired to work harder):

My self-esteem was at its lowest. The pianistic exuberance and the technical ease of Vladimir Horowitz made me feel deeply ashamed of my persistent negligence and laziness in bringing to life all the possibilities of my natural musical gifts. I knew that I had it in me to give a better account of the many works which I played in concerts with so much love and yet with so much tolerance for my own lack of respect and care.

Later on in the memoirs, early hints can be seen that he had 'gone off' Horowitz, for example when he described him as 'the greatest pianist, but not a great musician.' Never one for practising very much (though he did begin to do so more later on), he recommended that young pianists should practise no more than about three hours a day. With artistes such as Horowitz making recordings, Rubinstein was aware that he'd got away with quite a lot in his earlier career. There had been wrong notes, glossed-over passages and off-days. He became more self-critical, particularly after the recording industry had reached a better technological level and he could get more out of the recording process and listening to his and other pianists' recordings. On the other hand, after becoming a 'seasoned pianist celebrity', he felt that the younger generation of pianists were often technically proficient but at the expense of having little to say musically, saying that they play the piano 'too perfectly', and that he didn't play the piano as well as most pianists, nor did he work as hard. However, he added that when he heard these younger pianists, he had one little question for them: 'When will you start to make music?'

Before the Second World War, he married Nela in 1932 and would eventually become the proud father of five children (though one died in infancy): two daughters and two sons. His wife (shown right) was the daughter of Polish conductor Emil Mlynarski. Nela had first fallen in love with and married another Polish pianist, Mieczyslaw Munz. The marriage did not last very long and resulted in a divorce. Nela always said she had fallen in love with Rubinstein when aged eighteen. She'd gone backstage after a concert to meet him. To Rubinstein, meeting her on that first occasion was like a thunderbolt. He said he knew she was the only woman he wanted to marry.

Despite the age gap, the marriage ceremony, replete with photographers and many onlookers, took place at London's Caxton Hall. Divorce would never feature in the couple's long marriage despite Rubinstein having several affairs. At the age of ninety, nonetheless, Rubinstein left Nela for the thirty-three-year-old Annabelle Whitestone (who had collaborated on his memoirs).

Happy times after the Second World War saw the Rubinsteins being very much a part of the American celebrity scene (in the 1950s they had homes both in Paris and America). Family life in California was busy and fun (his friend Rachmaninoff lived nearby). While Mr and Mrs Rubinstein became quite famous for their parties and rubbed shoulders with Prince Rainier of Monaco, Cary Grant, Charles Laughton and others, the children, similarly, would have parties with Liza Minnelli and the offspring of Charlie Chaplin. Having passed his driving test in America, Rubinstein promptly went out and bought an almost new dazzling white Cadillac convertible.

Early on in his career, Rubinstein had recorded performances with RCA recordings, but he was also in the right place for movies, where it was either his piano skills that were dubbed over the actor's playing or it was he who had the leading role as the film was about him. He did not, however, forget his roots, he visited Israel and also helped fund-raise by giving numerous benefit concerts for both Jewish and Polish causes (the statue, right, is in his birthplace). While he worked very hard travelling and touring, even when at home in America he would

also work hard. In 1961 he gave ten concerts in one season at the Carnegie Hall.

At the age of eighty-seven, he gave a final concert in aid of the State of Israel in California. The following year saw his last concert in England. Held at London's Wigmore Hall, where he had first performed some seventy years earlier, it was to help raise funds for the concert hall. Remembering the Rubinstein of later decades, the American music critic Harold C Schonberg describes a typical concert platform entrance:

Without trying, he lets the audience immediately feel that it is facing a Presence. Look, my lord, it comes. Polite, agreeable, not gushing, he accepts the homage due him. He seats himself, and his nose points toward the stratosphere like the prow of a jet going upstairs. The audience waits, breathless. Rubinstein is in no hurry. He must compose himself; he must think of the opening piece; he must wait for the last cough to dissipate before he puts his hands on the keyboard. Suddenly the auditorium is filled with golden piano sound. A typical Rubinstein concert is under way.

He is shown in his glory days, right, receiving an Oscar from the actor Gregory Peck, but reaching his eighties, he could barely see the extremes of the keyboard, his increasing blindness forcing him to retire. Adhering to his motto of not giving in, he rather welcomed not having to perform any more as it allowed him to enjoy his other interests, such as art, wine and good food. He certainly wasn't ready to bow out of life; he lived on and actively into his ninety-fifth year. Arthur Rubinstein died peacefully in his sleep at his home in Geneva, Switzerland in 1982.

Dame Julia Myra Hess DBE (1890 – 1965)

Myra Hess was born into a Jewish family in South Hampstead. Although living within a fairly strict Jewish household, Hess wasn't always the conformist type. She would openly ride her bicycle on the Sabbath, although being asked not to. Later, defying the convention that refined Jewish ladies didn't smoke in public, she sometimes made a deliberate point of doing so (remaining unmarried, she was later known to refer to any piano she owned as 'my husband').

Myra Hess studied at the Royal Academy of Music, most notably under the well-known Tobias Matthay, who had been a pupil there himself and studied composition under Sir William Sterndale Bennett and Sir Arthur Sullivan. He rose through the ranks to become a professor (he is shown overleaf). Matthay would have a significant influence on how Hess approached piano technique and tone production, which was largely based on one of his books, *The Act of Touch in All its*

Tobias Matthay has a blue plaque at his former London NW3 home. It's an area with many musical connections as NW3 has 20 or so musical plaques, among them: Hans Keller, Delius and Sirs Bliss, Elgar, Walton and Henry Wood.

Diversity (he later broke away from the RAM and launched his own piano school in Oxford Street). Clifford Curzon, Moura Lympany and numerous other pianists all displayed the Tobias Matthay school of playing. But Matthay composed also, and as an accomplished pianist Myra Hess would play a one movement piano and orchestra work of his in 1922 at the Queen's Hall, with Matthay conducting.

For the beginning of her professional career, in 1908 Hess made her debut concert at the Proms (Liszt concerto in E flat) under conductor Sir Henry Wood, whom she found to be kind and helpful. Barely eighteen at the time, it must have been a nerve-racking experience; she could recall many years later being paid three guineas. Hess also worked under Sir Thomas Beecham (whose grandfather founded the famous Beecham Pills company near Liverpool), whom she found impossible: he was enthusiastic but very unreliable in organizing rehearsals and was also known to turn up without his music.

At another concert with Henry Wood and still young, she played the Mozart D minor concerto and took a 'wrong turn'. Somewhat mortified, she apologized profoundly to Wood after they left the platform, feeling it was the end of her career and that he would never engage her again, to which he replied, "It's nothing. You might have gone into a wrong movement then we would have had some fun."

At the outbreak of the Second World War, the National Gallery had been closed and all the pictures (but not necessarily the frames) removed to secret safe locations. In a bid to boost morale, Hess came up with the idea of having lunchtime concerts at the gallery. When she gave the first one (with many more later), she expected just some of her friends and possibly 40 or 50 others might turn up. Ten minutes before the concert was due to start, she was told there were 1000 people queueing (some would have to sit on the floor because there weren't enough chairs). These concerts went on to be held 5 days a week, every week for the next 6½ years. On one occasion she told the audience, "I've never played so much and practised so little."

The lunchtime concerts at the National Gallery's home in Trafalgar Square saw appreciative daily audiences from a wide cross-section of society (even members of the royal family), as the gallery's then director, Kenneth Clark, detailed in his book:

What sort of people were these who felt more hungry for music than for their lunches? All sorts. Young and old, smart and shabby, Tommies in uniform with their tin hats strapped on, old ladies with ear trumpets, music students, civil servants, office boys, busy public men, all sorts had come.

Myra Hess visited America forty times and became a firm favourite over there. Interestingly, the mother of jazz pianist and composer Dave Brubeck (born in America to Elizabeth Brubeck) had been given piano lessons by Hess in England as at one time she had wanted to become a concert pianist.

In 1961 Myra Hess suffered a stroke and had to give up public performances, her last being at the Royal Festival Hall of the same year. She died, aged seventy-five, just a few short years later after suffering a heart attack. Her 1927 Steinway model D grand was later donated to the Bishopsgate Institute in London. Recently fully restored, it is still used regularly, notably for Hess-inspired free lunchtime concerts.

Vladimir Samoylovich Horowitz (1903 – 1989)

Vladimir Horowitz was a Russian-American pianist who rose to the status of being regarded by many as one of the greatest pianists of all time (actually, at birth his surname was Gorowitz, but he changed it slightly when a young adult). He came on to the concert scene about two decades after Arthur Rubinstein, though bouts of illness (and possibly the side effects of medication he was taking) would force him out of the limelight for four distinct periods in his long career.

Born into a Jewish family living in Kyiv (English spelling Kiev, capital city of Ukraine and part of the Russian Empire), Horowitz was the youngest of four children born to Samuil and Sophia. They were quite comfortably off, his father was an electrical engineer and worked as a distributor of electrical motors for German manufacturers. At an amateur level, the family was musical. Sophia Horowitz played the piano and it was she who gave Horowitz his first lessons. Her daughter, Regina, played also, and later became a prominent pianist in her own right. One son, Michael, played the violin. Their father, in turn, was a supporter of the arts and an opportunity arose to take the young Horowitz to visit their uncle Alexander, who was a close friend of the Russian pianist and composer Scriabin.

Soon after, when aged around ten, Horowitz's talent would be spotted by Scriabin after playing for him. (It is said the composer had early on become fascinated by how a piano's action worked and even built some pianos of his own.)

Having entered the Kyiv Conservatoire in 1912, he continued his studies there until aged sixteen or seventeen. In 1920 he gave his first solo recital at the Conservatoire and followed it with a tour of Russia. This proved essential as the family had lost all their possessions in the Russian Revolution. He had not, in fact, seen himself as a concert pianist at this time, life as a composer had been his first ambition. Things were so tight at this time that it was said payment for his concerts came in the form of bread, butter and chocolate rather than roubles.

Despite highly successful concerts in Leningrad in 1924, in truth, his career prospects were stifled under the Communist regime he and his family endured in Soviet Russia. In December 1925, Horowitz emigrated to the West, where he studied under pianist Artur Schnabel in Berlin, but another motive for moving there was to settle permanently outside of Russia.

Horowitz later said that to help finance his first concerts, he stuffed American dollars and British pound notes into his shoes.

In 1926 Horowitz enjoyed success with concerts in Germany, France, Italy and Switzerland. The following year, he made his successful debut in America at the Carnegie Hall, where he played Tchaikovsky's First Piano Concerto under conductor Sir Thomas Beecham.

In 1933 Horowitz married Wanda, the daughter of conductor Arturo Toscanini, in Italy. Both Horowitz and Toscanini would regularly collaborate and work together on the concert platform. A daughter, Sonia, was born in 1934. Time would show that, with Horowitz often away on tours, and Wanda accompanying him, their daughter had a rather neglected childhood, being left in the care of nannies and governesses. As an adult, she suffered a serious motor-scooter accident in Italy; sadly, evidence suggests she may have later taken her own life when only in her very early forties. Even years after her daughter had died, Wanda would burst into tears at the mention of her name.

To a certain extent, Horowitz was rather like a child himself, with Wanda attempting to attend to his every need. He was a somewhat fussy eater. When on tour, Wanda was part of 'team asparagus', endeavouring to find the right vegetables and food because most hotels could not meet his requirements. A carefully selected choice of videos also had to be at his disposal (westerns being a favourite).

1936 would see the first of several absences from the concert platform for Horowitz. It is beyond the scope of this short biography, but in a nutshell he was undoubtedly vulnerable and had a troubled soul. Nervousness and a lack of confidence – despite some extrovert showmanship seen on the concert platform – combined with a homosexual leaning at a time when it was still illegal, must have added considerably to his depressive tendencies.

Unsurprisingly, the maestro became known for his many foibles and, at times, fearsome and unpredictable temper tantrums. We get a bit of an insider's view from the Steinway piano tuner tasked to tune both the piano and client (for many concerts, he would only perform on his own Steinway, especially relocated to the concert venue, and he would only do concerts at 4.00 pm on a Sunday afternoon). Tuner Franz Mohr compared Horowitz with another famous client of his, Arthur Rubinstein. Despite both being Jewish, coming from middle-class families, their families living under communist regimes, and both the highly talented Horowitz and Rubinstein being the youngest sibling in their respective families, in most other ways they were like chalk and cheese. Franz Mohr observes:

People really loved Arthur Rubinstein, because he loved people. Whenever people sought his autograph, on the street or on a plane or train, he would always stop and chat with them. With Horowitz it was quite different. Horowitz was extremely shy and afraid of people he didn't know. Once as Horowitz walked the few steps from the Bellevue Hotel in Philadelphia to the Academy of Music, with a whole bunch of us around him (Horowitz was always surrounded by friends), a person tried to stop him, saying, "Maestro it is good to see you. I am coming to your concert tonight." Horowitz very abruptly turned to walk around the man, saying only, "Good for you."

Not Rubinstein. He always had time for people. And he would really talk with them.

Mohr went on to add how Horowitz didn't like children and that Mohr had been advised by a Steinway staff member not to bring his son with him when making a particular trip to Washington to tune Horowitz's piano for a concert. (Horowitz himself had arranged piano lessons for his own daughter. It must have been intimidating for the teacher, as both her father and grandfather – the great conductor Toscanini – insisted on sitting nearby and watching the lesson!)

Horowitz's longest period of absence would be from 1953 to 1965. By this time, he had long since been an American citizen (and he did do some recordings). Just before these years, he and Wanda had temporarily separated. She was known to be quite a formidable character; while always Horowitz's greatest champion, she had a critical ear and Horowitz valued her opinions. She comes across as being somewhat 'old school', with what she would have viewed as high standards. One interviewer, Bryce Morrison, sent across from England and encouraged by Horowitz's agent to get his client inspired to tour again (and to lighten Horowitz's negative thoughts about flying), later recalled his rare opportunity to talk with the great man:

Arriving at the Horowitz household breathless but on time (my watch had stopped) and fortunately wearing a tie (Wanda Horowitz, the pianist's wife and the daughter of Toscanini, harassed and formidable, had been known to turn those away who had the temerity to arrive late or who were unsuitably clothed, addressing them in no uncertain manner with 'you insult my husband'.) I was filled with trepidation not because of the pianist – who when caught in the right mood could be affable, sweetness and light itself – but filled with awe at the thought of his actual playing. True to form his appearance was disarming, all smiles and with his signature brightly-coloured floppy bow tie. He exclaimed in wonder over the length of my journey from London but catching his agent's eye, I explained that it was remarkably short, that the time literally flew by, and received a thumbs up sign from Harold standing discreetly behind Horowitz. A good start, then, to tempt a pianist who had become restless, tired of hearing about 'other' pianists and who wished to be considered an 'international' rather than a 'national treasure'.

There were mischievous side-long references to other pianists, to Rubinstein ('why does he keep playing when he is so old, I hope I can still walk when I am his age. Incidentally, once I started playing again he stopped coming to see me').

It was Rubinstein who voiced the opinion that Wanda was very cold and stern, possibly being a hard task master with Volodya – the informal of Vladimir, which friends and family used. Wanda's nickname for her husband was Pinci. Rubinstein suggested that it was partly Wanda's temperament and treatment of her husband that led to his breakdowns.

There are those who believe the great Horowitz had and still has a certain marmite label attached to him. Adored by his many fans, who would cheer, stamp their feet and demand encore after encore, others have found fault with either his repertoire or pianism (sometimes both). Both the late Glenn Gould and Alfred Brendel are two such contenders. Another, composer and critic Virgil Thomson, called Horowitz 'A master of distortion and exaggeration.' Mind you, Thomson got rather short shrift when found out at a Toscanini concert. Wanda noticed he'd been asleep during the concert (and was aware of his numerous negative reviews of her husband), so remarked: "I am Wanda Toscanini Horowitz, and I saw you sleep from the first note to the last. I hope you enjoyed the performance."

Other critics have blown hot and cold. Even entries in the *Grove Dictionary of Music* altered over the years, one minute being complimentary, the next more critical.

In 1968 he made his first televised concert, but the following year would see another absence, which lasted for five years. He spent this absence almost as a recluse, but it also saw him undergo convulsive shock treatment. He would later try medication again, but the side effects proved to be too damaging. They affected both his memory and speech (he later refused to take them). Horowitz toured America in 1974, and in 1982 he returned to Europe – the first time since the Second World War – and gave two concerts at the Royal Festival Hall. Tours in Japan, the following year, sadly, were not well received and prompted his final (shorter) absence.

He did tours in Paris and Milan in 1985, and also made the film *The Last Romantic* at his New York home. This was followed by a historic USSR tour where he played concerts in Moscow and Leningrad. He was also able to visit Scriabin's house.

1987 would see a tour of Europe, including a return visit to Vienna, where he filmed

Horowitz Plays Mozart. He was now in his mid-eighties but appeared to be remarkably fit and well. Two years later he started some recording sessions in October but died on 5 November of what was thought to be a heart attack. His friend, pianist Murray Perahia, had been at his home at the time. He was buried in the Toscanini family vault in Milan. I'll leave Horowitz the pianistic giant to provide some final words and food for thought:

Piano playing consists of common sense, heart and technical resources. All three should be equally developed. Without common sense you are a fiasco, without technique an amateur, without heart a machine.

Below: Skilful hands! When in his teens, his father altered his son's age by a year so that he could avoid military service and the risk of damaging his hands.

Horowitz and wife, Wanda.

Wanda would live until the age of ninety. A friend of Woody Allen, Wanda even had a small speaking part in one of his films. Buried with her husband in the Toscanini family vault in Milan, her coffin was actually broken open by vandals in 2004, possibly in search of jewellery.

Dame Fanny Waterman DBE (1920 – 2020)

British pianist and academic piano teacher Dame Fanny Waterman was born in Leeds to Mary and Myer Waterman. Her father was a Russian Jew who had emigrated to England to work as a jeweller. As a young child, she recalled the thrill of hearing Rachmaninoff at a concert in Leeds Town Hall.

After attending Allerton High School, Fanny Waterman won a scholarship to the Royal College of Music (she studied with Cyril Smith, though was also influenced by Tobias Matthay). She shortly after appeared at the Proms in 1941 as one of the soloists for Bach's Concerto in C major for three harpsichords, conducted by Sir Adrian Boult.

Her musical career would be interrupted by the Second World War and she turned more to teaching after marrying Geoffrey de Keyser, a doctor, and producing two sons. With Roslyn Lyons and pianist friend Marion, Countess of Harewood (later Marion Thorpe after marrying the MP Jeremy Thorpe), she cofounded the Leeds International Piano Competition. Her husband had initially said it would never succeed in Leeds, but she stuck to her guns and it grew into a competition of high repute (her husband, it has to be said, was always very supportive of his wife's career). The competition went some way in helping to kick-start the careers of numerous concert pianists of note, including Murray Perahia and Mitsuko Uchida.

Having a forthright personality, she was almost proud to be known in some circles as 'Field Marshal Fanny'; she blamed mobile phones and the popularity of electric keyboards for distracting children from their practice. She also saw in some students a general lack of 'knuckling down' and having a rigorous, disciplined approach to their practice and attitude when wishing to become accomplished professionals. Someone who perhaps had the right spirit was Clifford Curzon, who told her that he had once had eight lessons to perfect a single Beethoven chord.

Dame Fanny Waterman lived a long life dedicated to music and was approaching the age of one hundred and one when she passed away.

Alfred Brendel KBE (born 1931)

Tall and with a slightly more pronounced stoop after reaching his ninth decade, the much loved and admired pianist Alfred Brendel was once described as having the appearance of a sad clown.

Born in Moravia, Czechoslovakia in 1931, Brendel would be the only child of loving though largely unmusical parents. They moved to Zagreb, Croatia (formerly part of Yugoslavia) when the young Alfred was just three years of age and where his father ran a cinema. He began piano lessons at the age of six, but the family moved to Vienna, Austria, where he studied piano and composition at the Graz Conservatory.

With the Second World War unfolding, Brendel was briefly sent back to Yugoslavia to dig trenches. Once, when he had returned to Graz, he did in fact witness Hitler visiting the town. Known to be something of a sceptic, he later reflected on this experience and commented, "It was my first impression of mass hysteria. I saw the eyes of the believers and it inoculated me against belief of all kinds."

Brendel has always had wide interests and pastimes; to begin with he didn't see himself as a goal-driven musician seeking to have a professional career as a concert pianist, he always reflected that the not necessarily intentional 'slow burner' approach to his musical development and career proved to serve him rather well. Throughout most of his career, he has also had a keen interest in painting, literature and composition. Indeed, at his first public recital in Graz, at the age of seventeen, the repertoire included one of his own sonatas.

Although largely self-taught after the age of sixteen (though he did have some support from Swiss pianist Edwin Fischer and says that he was inspired by acclaimed pianists Alfred Cortot and Wilhelm Kempff), while living and studying in Vienna he would go on to become the house pianist at Vox recording company (at the age of twenty-one he recorded Liszt's *Weihnachtsbaum*, the work's première recording).

Success in 1949, after winning a fourth place in the prestigious Busoni Piano Competition, led to tours in Europe and South America. While his initial preferences had been for Beethoven and Liszt, the works of Mozart and Schubert soon followed. When playing for concerts or on tour, he would practise on average six hours a day (he has always aimed to practise the concert repertoire and has never been a fan of doing piano exercises).

A major breakthrough came for the pianist after giving a Beethoven recital at London's Queen Elizabeth Hall. A day later, his agent was approached by three record labels. It was around this time, 1971-72, when Brendel made a permanent home for himself in Hampstead, London (actually in the same neck of woods as earlier concert pianist Dame Myra Hess).

Brendel's first marriage had been to Iris Hermann-Gonzala, an Argentinian singer. They had one daughter, Doris. The name seems slightly out of tune with her adult occupation, for Doris Brendel continues her career today as a pop rock musician. Brendel would have three children by his second wife, Irene Semler: two daughters, Katherine and Sophie; one son, Adrian, a cellist.

Brendel became known for having the slightly peculiar habit of wearing plasters whenever he played. He has large hands – can stretch twelve notes – but the plasters are more for protecting his nails rather than his fingertips and he has said on more than one occasion that he would feel lost without them.

English poet and novelist Al Alvarez, a good friend of Brendel, viewed the pianist as having a somewhat anarchic spirit and observed that he enjoys philosophy, collecting cartoons, and writing wittily surrealistic poetry.

Having reached the age of seventy-seven, in 2008 Brendel chose to play Mozart's ninth piano concerto in Vienna for his final public concert. His audiences during his formative years had been slightly weary of him, there didn't seem to be a lot of warmth on either side. Over time, however, they grew to both understand him and love his performances, with the result that many encores almost became the norm. At his final concert, under the baton of Sir Charles Mackerras, the enraptured audience rose to their feet even before the finish. Described as 'the greatest living interpreter of Mozart' by *Guardian* journalist Ed Vulliamy, he described the concert's poignant and memorable conclusion:

Each city of music has its distinctive final embrace that refuses to let a loved one go. Had this been Moscow, we would have stamped; in London, a veritable rose garden would have landed on stage. In Vienna, the hardest place to win a standing ovation, the Austrians rose and applauded, and simply continued to do so.

Glenn Herbert Gould (1932 – 1982)

One might think the pianist Horowitz was eccentric until coming across Canadian Glenn Gould. The family name had been Gold but, unlike Horowitz, they weren't Jewish so in 1939 the family changed the name to avoid any confusion or possible antisemitism. Interestingly, another link – musical – is with Gould's mother's maiden name, Greig. Although the Norwegian composer spelt it *i* before *e*, the grandfather of Gould's mother was a first cousin of Edvard Grieg.

Back to those eccentricities for a moment, of which there were many. Throughout much of his lifetime Gould was in the habit of quietly humming to himself and drumming his fingers on any surface available (even during concerts and recordings he would sometimes conduct, sing and hum; he once jokingly came into the recording studio wearing a gas mask as a 'humming deterrent'). An almost admitted hypochondriac, he felt the cold acutely, would rarely shake hands or have any physical contact with anyone, and for concerts and recordings was ultra-fastidious about the position and height of the piano, not to mention carrying his very own foldable and rather worn-out chair around to use wherever he played.

He would be a child prodigy, complete with perfect pitch and able to read music before he could read words in a book. Born in Toronto to Russell and Florence Gold, the family was of Presbyterian, Scottish and Norwegian ancestry. Both parents were musical. Dad was a successful furrier who enjoyed playing the violin; Gould's mother Florence played the organ and piano. It was she who gave their only child – whom she had rather late in life, when aged 42 – his first piano lessons.

During her pregnancy she had deliberately exposed her unborn child to music. It was noticed that when still very tiny, their son would hum rather than cry, and wiggle his fingers as if trying to play chords. When not much older, he began to play his own pieces to family and friends. When aged six he first heard the famous Josef Hofmann play at a live performance and was truly captivated, later recalling:

It was, I think, his last performance in Toronto, and it was a staggering impression. The only thing I can really remember is that, when I was being brought home in a car, I was in that wonderful state of half-awakeness in which you hear all sorts of incredible sounds going through your mind. They were all orchestral sounds, but I was playing them all, and suddenly I was Hofmann. I was enchanted.

At ten, Gould attended the Toronto (later Royal) Conservatory of Music. In addition to studying the organ and music theory, he studied the piano under Chilean pianist and composer Alberto Guerrero. He passed his final Conservatory exam in piano at twelve, achieving the highest marks of any candidate. A year later he passed the written theory exam. Normally these would have been sat by people older than himself. Then and later, he had a phenomenal memory and said he required little piano practice, being able to carry all his music in his head.

'Boy, Age 12, Shows Genius as Organist' ran a headline for Gould's first public performance in 1945. It was followed by a public performance with an orchestra when he played the first movement of Beethoven's fourth piano concerto with the Toronto Symphony Orchestra. Two years later he gave his first solo recital; this was followed in 1950 with a CBC radio recital. Success continued in the fifties, with the rising star giving concerts and broadcasts across Canada (he also composed a number of works). Such public performances proved to be rather short term, but he did undertake a tour of the Soviet Union in 1957. He made his Boston debut the following year but by the age of 31 he'd grown to hate all concert work, feeling concerts were outdated and had no future. Consequently, his concert appearances were small in number, fewer than 200 during his entire career. But he was destined to live a short life – he died very soon after his fiftieth birthday. Horowitz was born around 30 years before him but would outlive him; rather oddly, Gould had told close friends he didn't foresee his life lasting beyond the age of 50.

He became famous for his brilliance in technique and interpretation of the works of JS Bach, notably the *Goldberg Variations* (his recording of this became a phenomenal hit). Casting public performances aside (concerts were 'an evil blood sport'), he felt really at home in the recording studio, saying he had a love affair with the microphone. Apart from writing and broadcasting, he would devote most of his time to the studio, revelling in its technical and more secluded environment.

The chair Gould used, be it for concerts or in the recording studio, had been a chair used for bridge games, which his father adapted for his son. The appearances of this slightly odd and, in time, battered 'pygmy' chair were almost as famous as its owner. When young, Gould had originally been encouraged to use the chair using a straight back, this was partly with the aim of reducing the chances of back ache.

There have been suggestions that Gould may have had Asperger syndrome, or that he was bipolar. He went through a stage of taking a lot of anti-depressants. He would be taken from the musical world sadly early. After suffering a stroke, which left him paralyzed on his left side, he was admitted to Toronto General Hospital. His condition deteriorated and it was evident he had suffered brain damage. His father decided he should be taken off life support. A public funeral took place in St. Patrick's Anglican Church, the service was attended by 3000 people and broadcast on CBC. He was buried in Toronto's Mount Pleasant Cemetery alongside his mother (and joined by his father after his death in 1996). He left half his estate to the Salvation Army and half to Toronto Humane Society, an animal welfare charity.

Despite the eccentricities – having a pet skunk, wearing a hat, gloves and overcoat even during warm summer days, or eating the same (English) meal at the same Chinese restaurant, in the same seating area and at the same time each day – the tribute by Jed Distler in *Gramophone* magazine below helps to explain why he has legions of fans long after his death:

Glenn Gould threw received notions of piano interpretation and core repertoire out the window, while his unique hair-trigger articulation, particular sonority (two notes and you know it's Gould), rhythmic focus and X-ray-like contrapuntal clarity left music lovers mesmerised and his baffled colleagues awestruck. His recordings and television performances alternate between revelatory and infuriating, yet they consistently entertain and beckon your attention.

Thirty years after his death, a recording of Gould playing Bach's Prelude and Fugue in C was included on the spacecraft Voyager 1's 'Golden Records' that were sent into interstellar space in the hope that extra-terrestrials might one day hear it.

Below: A youthful Gould with Mozart, the parakeet. Right: The pianist with his famously low chair. His

preparation before any concert was to soak and massage his hands and forearms in warm water for twenty minutes. After his death, the Glenn Gould concert hall and recording studio was built and came complete with a bronze statue outside.

PIANISTS OFF-DUTY... CAN YOU RECOGNIZE WHO THEY ARE?

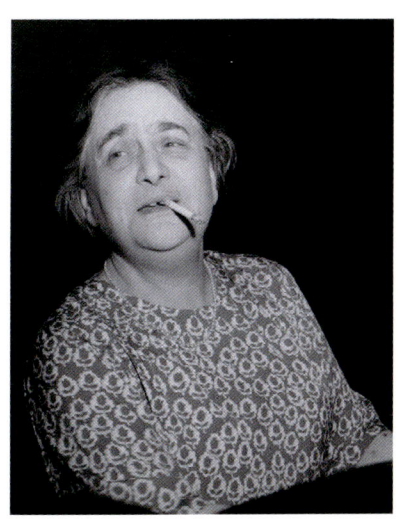

FAMOUS PIANOS (SELECTED EXAMPLES OF THE MANY)

It is not my intention to list every famous piano that still exists, but just to give a light perusal of some that are either famous or interesting in some way. A few were not necessarily wonderful instruments but were made famous by where they were used or by whom. Captain Scott's ultimately tragic endeavour to get to the South Pole, for example, saw the use of a piano as being important for the team in regard to morale and well-being. A Broadwood upright pianola was taken and did sterling service, meeting the needs of non-pianists and pianists alike.

The two Steinway pianos belonging to jazz pianist, Fats Domino, had rather ignoble ends after suffering severely through the storms of Hurricane Katrina (though one was restored and the other made an unusual exhibit in an art exhibition).

Quite often, famous performers owned more than one piano; certain other instruments have had claims to fame, the often-ordinary piano having been used by someone famous. In some cases, the claims that someone famous once owned the said piano, or composed a piece of music on it, have proved to be apocryphal (and possibly motivated with the intention of adding kudos and value to the instrument).

A rather forgotten grand, at one time in great demand by pop musicians, is the 'Trident Bechstein' (see overleaf). Thought to be around a hundred years old, it was rented long-term from Jaques Samuel Pianos in London's Edgware Road. In some ways, this grand was a bit of a beast, its action was heavy and its tone on the harsh side. It wouldn't have cut the mustard for most concert pianists, yet at its home in the well-known Trident recording studio, it became quite legendary. From the late sixties to mid-eighties, the Bechstein was used for many iconic recordings, largely because of its unique tone. It can be heard on (to name just a few of the many) recordings of The Beatles' *Hey Jude,* Elton John's *Your Song*, the Boomtown Rats' *I Don't Like Mondays*, and Carly Simon's *You're So Vain*. David Bowie also used the piano, as did Freddie Mercury. When the Trident studio closed; it is said the Bechstein was auctioned off (despite being dropped during the removal).

An Eavestaff minipiano owned by Alma Cogan. It became known as the 'Yesterday Piano' after Paul McCartney visited the singer and played the song (he was convinced the tune had already been written by someone else so wanted confirmation that it hadn't been). The working title for *Yesterday* was *Scrambled Eggs*. McCartney recalled recently his early inspiration for learning the piano: *Music was a big thing in our house. Our family's annual New Year's Eve party was a joyous thing where everyone sat around singing. Dad was on the piano banging out these old songs that everyone knew...*

Left: John and Ono Lennon's Steinway grand. It was a brand-new model finished in mahogany, but John wouldn't take delivery until it was sent away and refinished in white.

Left: The Trident Bechstein in situ at the Soho studio; it was an old but robust piano with character.

Right: Elvis Presley's Kimball gold piano, which was put on display at the Hard Rock Hotel, USA. He also owned a second, white 1912 Knabe grand which he kept at *Gracelands*. The piano, right, wasn't originally gold, it had been bought for Elvis's mother. After her death, Priscilla Presley had it all gold-leafed and presented it to Elvis as a first anniversary wedding present.

Left: The White House art-case Steinway grand.

Presented to President Franklin D Roosevelt, the White House Steinway grand was made in 1938 and was the company's 300,000[th] instrument.

Left: Dog handler, Cecil Meares, in the hut set up on the ice at Ross Island during Captain Scott's sad and fateful Antarctic race for the South Pole. Captain Scott personally chose the piano while visiting the Ideal Home Exhibition. An earlier expedition of 1901-04 had shown that most of the men could not play the piano, so it was thought prudent to supply a pianola (it was made by the long-established John Broadwood company). However, Scott did appreciate the pianistic skills of one of his team on the earlier expedition, that of Lieutenant Charles Royds, who would play the piano before the evening meal each day. Scott wrote in his diary: *His hour of music has become an institution which none of us would willingly forego*. The piano played by Royds went into his family and is currently owned by his grandson, the film and theatre director Richard Eyre. Perhaps a precedent for pianos on expeditions had already been set, as a piano was also taken aboard the *Discovery* by Captain George Nares on his British Arctic expedition in 1875.

Right: Jazz performer Arthur 'Dooley' Wilson playing Sam alongside Humphrey Bogart in the film *Casablanca* (1942).

Left: The Casablanca piano was a studio prop and only had 54 keys. It was sold by auctioneers Bonhams in 2014 for $3.4m.

Popular Trinidadian ragtime/honky-tonk pianist Winifred Atwell first came to prominence in the 1950s. She had had some classical training and gained a place at the Royal Academy of Music, but made a name for herself after appearing on TV's *Stars in Their Eyes* in 1946. It was her battered old upright, coined by Atwell as 'my other piano', which became almost as famous as her; it is now owned by Sir Richard Stilgoe (there were in fact two of these 'My Other Pianos'). Atwell's first piano was bought from a junk shop in Battersea for 50 shillings, it

travelled thousands of miles on her world tours (she and her manager-husband settled in Australia). Atwell had two number one hits in the UK and her 1952 recording of *Black and White Rag* became the theme tune to the BBC's television snooker programme *Pot Black*.

Below left: Former prime minister Edward Heath seated at his Steinway, which he bought with the £450 he won in the Charlemagne Prize for leading Britain into the EEC. It was moved into 10 Downing Street and, in his later years, to his home – *Arundells* – in Cathedral Close, Salisbury. *Arundells* is an early eighteenth-century country house; after prime minister Heath's death, his former home was opened to the public under the Edward Heath Charity. The home and piano are still used for private functions and concerts.

Heath's Steinway grand is in contrast to his first piano, bought for him as a nineth birthday present (his parents paid £42.00 on HP for the Thornton Bobby upright).

Right: One of concert pianist Paderewski's pianos, this one (a Steinway) is in the Polish Embassy, USA. His fame stretched out in many directions during his lifetime and after, including being mentioned in the Irving Berlin song *I Love a Piano*: 'And with the pedal, I love to meddle/When Paderewski comes this way/I'm so delighted, when I'm invited/To hear that long-haired genius play.'

Left and below: Two grands that belonged to jazz performer Fats Domino. Both, sadly, were severely damaged when Hurricane Katrina hit New Orleans. The top one was put on display as part of an art exhibition, the white Steinway was later renovated after a restoration fund was launched (Sir Paul McCartney was one of the numerous donors). The grand had been flipped over during the hurricane and stood in ten feet of water.

Below: The Victorian artist and designer Edward Burne-Jones' Priestley piano, given to him as a wedding gift (he is shown below, right). The small straight-strung upright has only bichords (two strings per note) throughout and stands three feet high. Its walnut case was left plain and unpolished so that he could paint the reliefs directly on to the case.

Left: The Royal Suite in London's Claridge's Hotel, which first opened in 1856. They have several pianos but this grand (a Broadwood) was originally in the ownership of Richard D'Oyly Carte, who in addition to building the Savoy Hotel and Theatre, founded the D'Oyly Carte Opera Company, which championed the Gilbert and Sullivan operettas (the Savoy Theatre was the first public building in the world to be entirely lit by electricity). D'Oyly Carte married Blanche Prowse, the daughter of William Prowse, a piano manufacturer and music publisher. A night's stay in the Royal Suite starts from around £6000 (butler included).

Right: Alicia Keys' Yamaha grand, which as well as decorating it herself, she often took on tour with her.

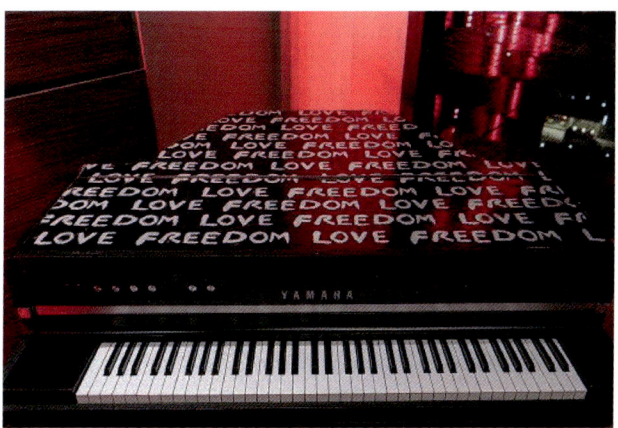

The Steinway Model Z that belonged to John Lennon (on which he composed his song *Imagine*). When it came up for auction in 2000, the pop singer George Michael bought it for £1.67 million and arranged for the piano to be part of Liverpool's The Beatles museum's collection.

FURNITURE (BOTH INSIDE AND GARDEN)

To some, it may seem that the piano is dead, but in keeping with trendy thoughts, the terms recycled, upcycled and repurposed, are very apt. The sight of a piano being dismembered and cast aside (butchered and sent to the slaughterhouse, but I mustn't exaggerate), to me, is rather heartbreaking, and yet one can have admiration for the many creative ways some unwanted pianos are being kept alive, even when not being used for their original purpose.

With the garden features on these pages, we mustn't forget that there are plants named piano too. Above are roses known as the Pink Piano Rose.

There's music in water, if you stop to listen...

Left: Lastly, perhaps piano furniture is not such a new idea. This small keyboard was given to composer Franz Liszt by the Bösendorfer piano company. Eagle-eyed readers might also spot two tuning forks in front of the framed photograph. The desk is on display at Budapest's Franz Liszt Memorial Museum.

G

GUINNESS; GRANDS; GIRAFFE PIANOS; GRAVES

A Guinness World Record! On 4 March 2017 in Lisbon, Domingos-Antonio Gomes, a Portuguese-American pianist, smashed the record for the 'Most piano key hits in one minute', proving he has the fastest fingers in the business. The pianist played a note on a Yamaha grand a mind-blowing 824 times in 60 seconds – that's more than 13 hits a second! (The note played was B7, which is next to the piano's highest note, top C; the first/lowest B on the piano is known as B0 not B1.) In order to beat the previous record of 765 hits, Gomes practised his technique for four months, which involved alternating between two fingers to press the key as well as using a metronome to keep a steady rhythm.

The largest piano in the world (and another Guinness World Record), one that actually fully works as a piano, measured 2.495 m (8 ft 2 in) in width, 6.07 m (19 ft 10 in) in length and 1.925 m (6 ft 3 in) in height and was constructed by Daniel Czapiewski. It was played in a concert at Szymbark, Poland, on 30 December 2010. In fact, it contained 156 keys, and two grand pianos could be placed side by side underneath it.

Photo overleaf: You can hit a long note rather than a wrong note on New Zealand piano tuner Adrian Mann's extra-long piano. Roughly twice the length of a concert hall grand, it was a world record in 2009, measuring 5.7 m (18 ft 9 in) in length. Built on his farm over four years, it required the help of the local fire brigade to move it when put on display.

Its maker actually started building the piano when aged only fifteen. The frame is made from welded steel; numerous pianists of note have visited and played the piano, wanting to savour its unique tone in the bass.

GRANDS

There weren't grand pianos to begin with, the first common pianos seen almost everywhere were flat and usually had four legs, not three. They were known as square pianos but were actually narrow and rectangular in shape. When the piano evolved and they became both 'beefier' in size and with a wider compass, the shape and design of the piano developed into the larger wooden-framed fortepiano (like a harpsichord in appearance) and then the iron-framed grand piano. The term 'grand' started to get some usage – perhaps also along with the terminology and concept of a 'Grand Concert' – as the piano gradually started to be seen as an important solo instrument with an increasingly wide repertoire being composed for it. The first grand pianos to appear in England were made by Americus Backers from the 1770s. He lived in Westminster and advertised his new invention (though it wasn't quite that and hadn't been patented) in *The Public Advertiser* of 1771:

In the Long Room in the Thatched House, St James's Street, on Monday, Tuesday, Friday and Saturday Mornings, may be seen and heard a new invented Instrument of the Size and Shape of a Harpsichord, which answers all the Purposes that have been hitherto wanted in an Instrument...

It was stated that a fine harpsichord player was engaged to play it between one and two, but otherwise any lady or gentleman could try the instrument (admittance 2s 6d). The firms Stodart and Broadwood followed by launching their own grands (Robert Stodart was the first maker to use 'Grand' when patenting his instrument).

On most good grands, because of the design of the action (compared to an upright action), it allows the pianist to play faster. Ranging from around four feet seven to five feet seven in length, a baby grand is the smallest type of grand. A slightly larger model would give a better tone due to its larger soundboard and longer string-length, this kind often being referred to as a boudoir (or parlour) grand, being around six feet in length. Pianos over six feet in length can be called a small (or semi) concert grand, being suitable for recital work and concerts in smaller venues. Larger and more impressive concert grands, especially needed for piano concerti, are from seven to nine feet in length (piano types and classifications can differ both between makers and retailers in the UK and overseas).

Are grand pianos elitist? Perhaps not unless you get into the world of comparing, say, a Bösendorfer with a Steinway. In the recent austere times, a row surfaced about a Sheffield college's 'grand ambitions', as reported in the *Sheffield Tab*:

University of Sheffield bosses have been slammed by their staff over the decision to splash the cash on 17 prestigious Steinway pianos, worth £472,000, for the music department. One staff member told The Tab *that the decision was purely a 'display of elitism' and an 'obtuse gesture made by university executives who are completely out of touch with staff and students'.*

The decision was made pre-COVID, and staff were fuming at the time over the waste of money – but with crippling losses from the pandemic, rage has heightened and questions have been asked about why the purchase is still going ahead.

"a shocking waste of money"

The university said that the pianos will 'enhance the student experience', and that they remain contractually committed to the 2019 purchase. However, amidst sweeping job cuts and a university tutor warning of devastation to student learning, it's been pointed out that keeping hold of valuable staff will benefit students more than a few pianos.

Mozart wrote a piano concerto for three pianos, but American minimalist composer Steve Reich wrote something that required six pianos. The original concept for his work had been a piece first titled *Piano Store* and was intended for musicians to

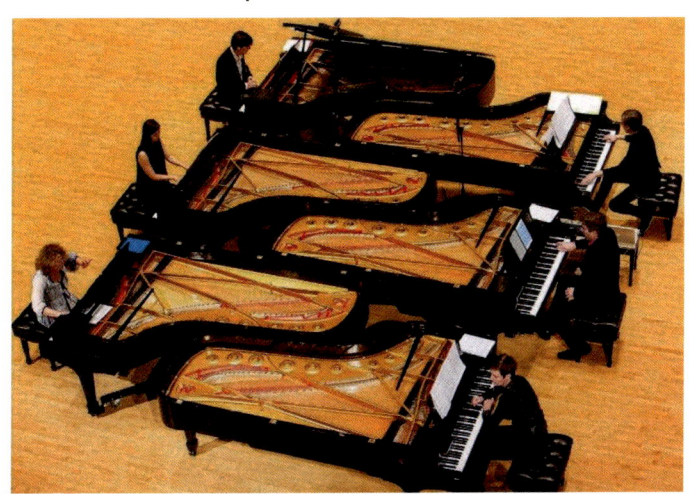

play on all the pianos in a piano store. As a work for six grand pianos, it had its New York première in 1973; it has since been performed around the world. The New York based music group *Grand Band* (shown left) was formed in 2014. They performed it, but then realised it was worth staying together, having found a demand for numerous other works of a similar genre.

GIRAFFE PIANOS

19th century Giraffe (or Giraffenflügel) upright pianos were designed to take the strings quite high up in order to get sufficient string length and tension (but must have been difficult to reach and tune). It took a while for makers to think about starting the strings from ground level rather than keyboard height to avoid the instrument being too tall.

Left above: An American Teupe Giraffe piano. Above right: 'Music, When Soft Voices Die, Vibrate in the Memory' – a painting by the noted Scottish painter Sir William Quiller Orchardson (1832 – 1910); pianos also feature in other works of his.

The reader will see later on that the invention of the English upright piano is generally credited to John Isaac Hawkins, but there had been other attempts which didn't catch on. A much earlier inventor and engineer is the Italian Domenico del Mela (also a priest, teacher and maker of harpsichords and organs). Alive at the same time the accepted inventor of the piano Bartolomeo Cristofori was still working, he moved the design concept of pianos being flat and table or harpsichord-like on to the instrument standing upright, being vertical and no longer horizontal. His 1739 upright piano (though not called that) is still in existence, being kept by the Museum of Musical Instruments in Florence; however, it was something of a one-off and did not quite lead to other piano makers building instruments along identical lines.

GRAVES

A rather forgotten piano maker is the Italian, Muzio Clementi (1752 – 1832). Known by some as 'Mozart's Rival', he became a London resident and very well known as a musical prodigy, composer and publisher. His Westminster Abbey grave epitaph clearly describes him as 'The Father of the Pianoforte', and he certainly both promoted the pianoforte and established a successful piano manufacturing business with F W Collard (which later became the firm Collard & Collard after family members eventually became sole owners, the family vault is in Highgate Cemetery).

For numerous other families of piano builders and musicians, it would seem they wanted to be creative in remembering their dearly departed loved ones. Included below and overleaf are both the famous and less well known.

The above grave can be found in London's Highgate Cemetery. The cemetery is famous for Karl Marx's grave, but also has other famous ones: Michael Faraday, George Eliot, and Douglas Adams. The Henry Thornton grave (above, he is depicted with his wife below) is

that of 'Harry Thornton', a classical pianist and entertainer who, with his wife, entertained the troops during the First World War. He succumbed to the flu epidemic in 1918.

A lyric on the side of the piano reads: 'Sweet thou art sleeping; Cradled on my heart; Safe in God's keeping; While I must weep apart.' The words are an English translation of lines from Puccini's opera *Madama Butterfly*.

Above: City of London Cemetery. Grave of Gladys Spencer LLCM, a Manor Park music teacher who died of pneumonia at the age of 34 in April 1931. She also ran a dance troupe of 'high kicking' young girls who put on shows. Part of the affectionate wording on the grave is 'to my Darling Gladeyes'.

Right: Cedar Hill Cemetery, Vicksburg, Mississippi.

Below: Grave of Dudley Moore CBE. Pianist, composer and entertainer, he at one time

owned three grand pianos. He is buried in Hillside Cemetery, Scotch Plains, New Jersey. The son of a Glaswegian railway electrician, Moore was brought up on an estate in Dagenham. When aged 11 he earned a Guildhall School of Music scholarship, the piano and organ would have to wait however, as it was here he first studied general music and the violin. He later won an organ scholarship to Magdalen College, Oxford.

Left: St. Peter's, Limpsfield, Surrey, grave of Australian pianist Eileen Joyce. She was known to enjoy playing on Bechsteins during her career and it is her who can be heard playing Sergei Rachmaninoff's haunting theme from his Second Piano Concerto in the 1945 film *Brief Encounter*.

Top and left above: The Berlin family grave of the founder of the Bechstein piano company.

Below: A memorial tablet in St. John's church, Crawley to the family of Britain's oldest firm of piano makers, John Broadwood & Sons Ltd. In 1761, John Broadwood had walked from his Scottish home to London as a young journeyman cabinet maker. He would leave his sons a highly successful piano manufacturing business and estate. Composers such as Haydn and Chopin visited the London showroom. One son, Thomas Broadwood (later

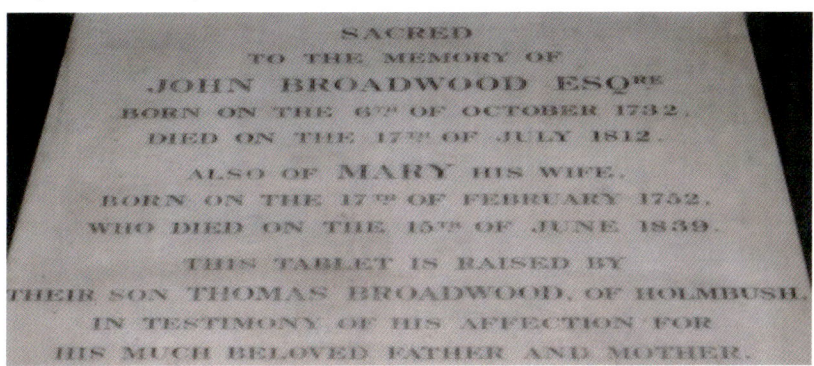

appointed Sheriff of Sussex), would visit Beethoven at his home in Vienna and hear him play. About a year later, in 1818, a new Broadwood grand was presented to the composer.

79

Above: A few bars from the *Goldberg Variations* can be seen on Gould's memorial tablet in Toronto.

Right and right above: An isolated grave of piano tuner Fred Schlieffen. Born in Germany, he lived and worked in Australia. His unfortunate demise was to die by drowning in floodwaters in 1906.

Below: We can't forget composers such as Schubert, who wrote some wonderful pieces for the piano.

Below: American jazz pianist and composer Joe Sample.

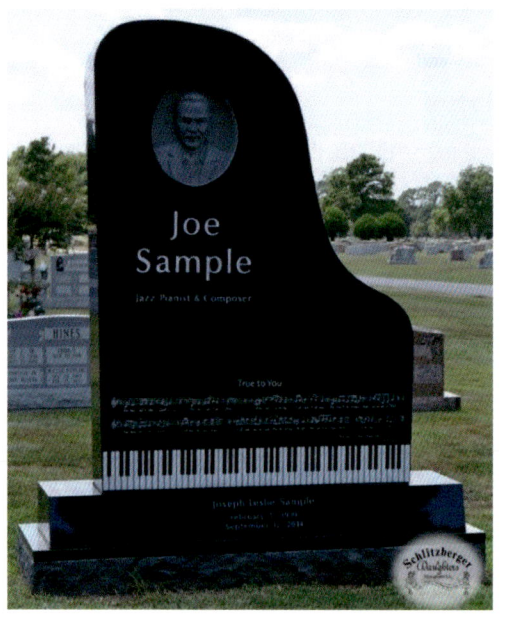

H

HORSES; HOTELS; HARRODS (AND DEPARTMENT STORES)

Horse racing has used piano-related names from time to time. *My Old Piano* (sadly recently deceased) was a racehorse from Yorkshire. A younger example is the Australian thoroughbred racehorse named *Grand Piano*. Born in 2016, its mother went under the name of *African Piano*. Returning to England, gelding *Pianoforte* won at Wolverhampton in 2009. (No horsing around on my part, back in 1906 there was a racehorse named *Tuning Fork*, with another UK thoroughbred named just the same this century.)

HOTELS

Restaurants styled as piano restaurants within certain hotels have come into vogue in recent years and there are numerous hotels around the world which have either 'piano' as part of their name or use a piano-themed décor. One example is the Piano Hotel in Izmir, Turkey, which has a piano theme inside and out. India is another example, it has The Grand Piano Hotel in Patan (shown overleaf).

Above: The Piano Hotel, Izmir, Turkey.

Below: Claridge's Grand Piano Suite, London. Said to be the royal family's favourite hotel, the hotel has numerous other pianos, which is not surprising when past regular temporary residents have included André Previn and Daniel Barenboim.

Another famous hotel nearby, The Dorchester, recently purchased Liberace's mirror-tiled Baldwin grand (the pianist had quite a collection of pianos). Completely renovated – including the 1000 or so tiles – it has pride of place in The Dorchester's new Artists' Bar.

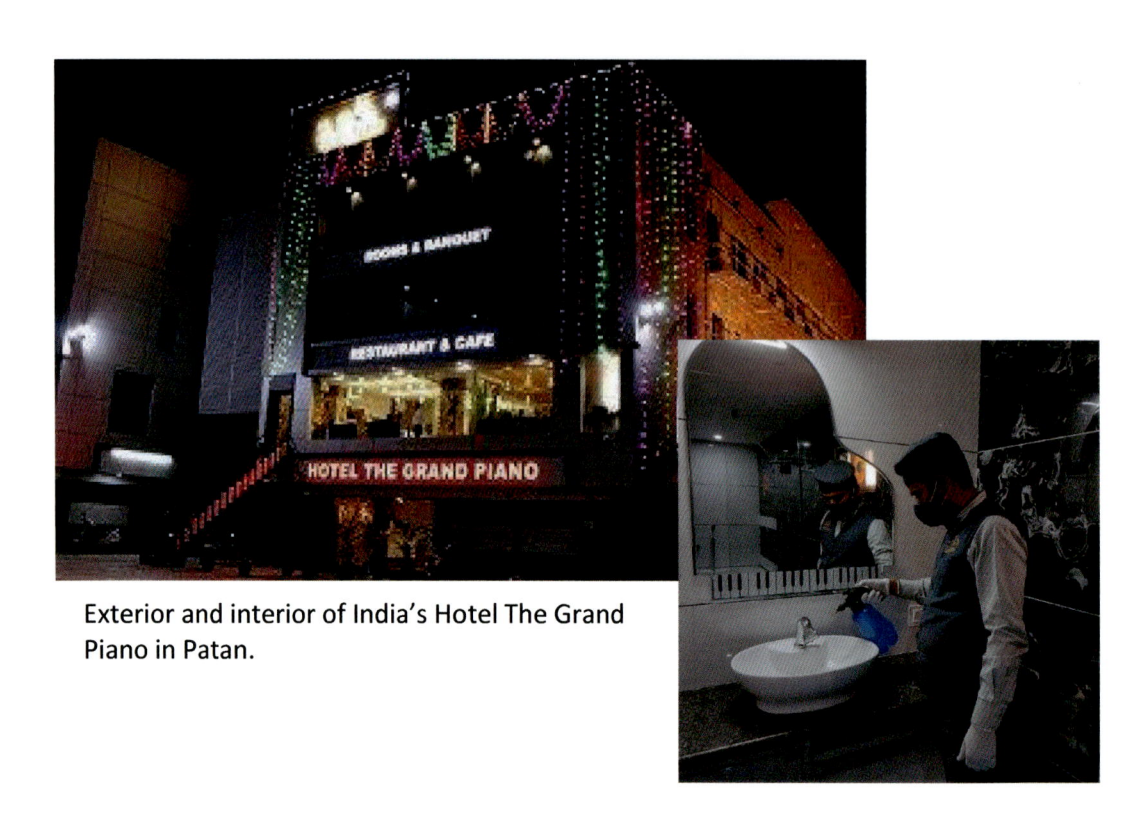

Exterior and interior of India's Hotel The Grand Piano in Patan.

HARRODS (AND OTHER DEPARTMENT STORES)

Harrods was one of the first, or possibly the first, department store to have a piano showroom (often called salon). The department having started in 1895, the press was quick to deploy appropriate headlines when it was eventually decided to close the department in 2013. 'Harrods closes lid on Pianos' was the type of headline that was published, along with subheadings about handbags being more popular than pianos. The store itself referred to the demise of the piano industry in the UK and that each year fewer and fewer people were buying new pianos. Just before closing, one new Bechstein was price-slashed from £94,000 to £66,000 (the store does occasionally still sell pianos).

How quickly we forget that, across the country, almost every major department store – be it John Lewis in London or John Lewis in Liverpool – once sold pianos. Barkers of Kensington, in 1924, opened branches in Birmingham, Manchester and Liverpool, each one included a piano department. When pianos of all kinds were mass produced and in high demand, department stores were able to cash in and take some of the profits from the specialist retailers. Indeed, certain stores had pianos made exclusively for them, with piano manufacturers making (rather plain) pianos that carried the store's own name (see the Gamage's piano overleaf – they never actually made their own pianos).

With these department stores, the world of discounts and related services such as tuning and repairs were all sewn up (though there was still sufficient trade for most of the specialist piano retailers to run two or more branches). No doubt a correlation would be seen if one were able to compare sales figures and dates between the twenties and thirties and fifties and sixties. Noël Coward, once his career finally started to blossom, proudly walked into Harrods and purchased a second-hand grand. His father had earlier been an unsuccessful piano salesman for Paynes Pianos, but for the young Coward it had to be Harrods (or Selfridges at a push). No, piano showrooms in the twenties and thirties had instruments packed in and lined up as though selling books; people walking into the same stores in the fifties and later were still drawn to the better, often attractive mahogany veneered cases, only now they came with on and off switches. The television took up much less space but from these and radiograms you had instant entertainment and music. Pianos as wedding presents were perhaps less fashionable after the war, but the grand could still be found in many middle-class homes. Several decades earlier, before widespread ownership of cars and electrical goods, it wasn't uncommon for pianos – considered valuable assets – to be stolen or to feature in custody and divorce cases. Others were repossessed when aspiring but poorer owners struggled to keep up with repayments under the popular hire purchase schemes.

Being harder to sell, Harrods had to think of new ideas to garner interest. They held weekly concerts where a well-known musician (Semprini, Liberace and others) would show off selected models. Even in the seventies, Harrods managed to get jazz pianist Oscar Peterson to perform; on another occasion they put on a concert which needed sixteen pianos being used at the same time. As I write, there are no department stores in the UK who sell pianos. The London John Lewis store, after their 2018 Christmas commercial was aired, which showed pajamas-clad boy Reg Dwight (Elton John) and adult performer Sir Elton John playing the Christmas wrapped present of an upright piano, received minor media flack. It was pointed out that the store didn't sell pianos and hadn't done so for many years. Was it simply damage limitation? John Lewis began advertising pianos for sale again (using a bit of licence – they were electric keyboards).

Right: Harrods' pianoforte showroom, 1930s. They were then quiet places and pianos could only be played with a suited gentleman on hand to discuss the merits of a Bechstein over a Blüthner (uprights were a somewhat pushed-to-the-side last resort option).

Pianoforte Salon

Left: Selfridges, Oxford Street. Opened in 1909, pianos became a fixed feature. Perhaps no one could foresee how television would change people's daily entertainment habits, except, possibly, John Logie Baird.

For one month in 1925 John Logie Baird demonstrated his moving silhouette images by television on the first floor of Selfridges. A slight irony, perhaps, is the fact that his small laboratory was in London's Soho, an area where several of the first leading English makers of pianos established their workshops and showrooms. (Actually, at one point there was a grand piano in the Baird family home, for his South African wife Margaret Albu had been a concert pianist!)

Above: Known for its extensive toy department, there was evidently also demand for and money in pianos. In

1924, a Gamage's advertisement ran: *The home of the world's best makes of pianos and gramophones.* Even well-known London furniture stores Maples, and Waring and Gillow (both suppliers of furniture to posh hotels and the royal yachts) sold pianos at this time.

Queens Road, Bristol (Duck, Son & Pinker Pianos, who had several branches). A major high street where you couldn't buy a piano? Almost unheard of before the Second World War.

Even after the Second World War, the investment of a piano was worthwhile for many. The Dwight family started with a basic King Brothers upright (made over Hackney way). Today Sir Elton John owns several grands. One of them cost him – or was known as – the million dollars piano, the said piano is still acoustic but comes with digital technology so that it can play and control numerous visual and aural effects on big screens. It might be a Yamaha, but to its owner it is called *Blossom* – named after a favourite singer of his.

Right: Sheet music and piano party/sing-song magazines were profitable in the 1950s. In pubs, boys' clubs, church halls, hospitals and for school assemblies or the Women's Institute, an upright piano was still a valued and very necessary item. Born in the late fifties, musician Jools Holland's family couldn't afford a £10 second-hand one at the time, but the young Jools could practise on his nan's piano though. He wrote in his autobiography: *Genetically, the fact that my two grandmothers, my mother and my uncle played the piano may have had something to do with my ability. And I loved playing more than anything … being introduced to the thrill of playing the piano changed the course of my life.*

I

INNER EAR; IVORY; INVENTOR (OF THE PIANO)

There's an inverse piano in your head! Neuroscientist James Hudspeth, interviewed in the journal *Scientific America*, recognizes well the claim, having researched the subject for several decades.

The nineteenth century German scientist Hermann Helmholtz recognized that the inner ear's cochlea is a kind of inverse piano. Pianos, as we know, have individual strings. Each one represents a single tone or they can be used together to make a harmonious whole new sound. The ear, apparently, undoes that work, taking the 'harmonious whole' and separating out the individual tones, representing each one of them at a different position along the cochlea spiral. Hudspeth tells us that each of the 16,000 hair cells that line the cochlea is a receptor that responds to a specific frequency. All the hairs are in a systematic order, just as the piano strings are:

The ear has a built-in amplifier, and that amplifier is unlike any of our other senses. It would be as if light going into the eye produced more light inside the eye, or smell going into the nose produced more smell molecules. In the case of our ears, the sound that goes into the ear is actually mechanically amplified by the ear, and the amplification is between 100 and 1,000-fold. It's quite profound. And the active process also sharpens the tuning of hearing, so that we can distinguish frequencies that are only about 0.1 percent apart. By comparison, two keys on a piano are 6 percent apart.

IVORY

White piano key tops were traditionally made from ivory, it was a sign that the piano was likely to be of good quality. Cheaper pianos used celluloid (which has been around since the 1860s), but even pianos with ivory tops had key fronts that were bone (in very early pianos), wooden or often of a plastic material, not ivory. The use of ivory key tops ceased on all pianos after a world-wide ban on the ivory trade in 1989, or even earlier in some countries (organ keyboards also used ivory).

With the piano keyboard traditionally having white naturals and black sharps/flats, this hasn't always been the case for pianos and harpsichords. Mozart would have been familiar with both instruments having the reversed colouring; possibly, when it was harder to source ivory and it was for a time more expensive, perhaps it made better sense to cover the fewer sharps and flats in ivory and use ebony or rosewood for the naturals.

Ideally, ivory was the best material to use because it was a natural material that had the right feel and touch under a pianist's fingers. Being porous, it was able to absorb perspiration and so was much less inclined to become slippery. For concert pianist Arthur Rubinstein, whether it was ivory or not, he wanted that certain degree of grip under his fingertips, so had two tricks of the trade. He would either apply a small amount of the rosin used by violinists to his fingers or, believe it or not, get the piano technician to lightly apply some hairspray over the keyboard.

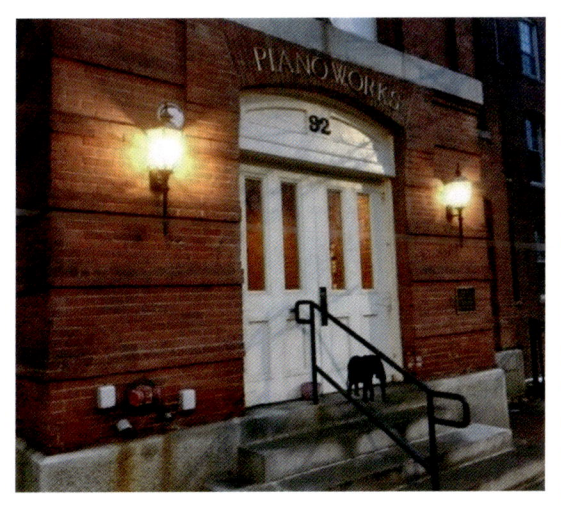

From c.1840 to 1940, the USA was the biggest importer of ivory, which resulted in thousands of elephants being killed in the process (whether for trade or trophy hunting, ivory became known as 'white gold' by some). In Connecticut, the piano trade ceased a long time ago, but the town of Deep River still has plenty to remind it of its past association with pianos, or more so ivory. Even before the widespread growth of pianos throughout most of the world, ivory was imported from Portuguese east Africa and taken upriver to loading docks a short distance from the town's factories. Here, ivory was used for making such things as billiard balls, combs, toothpicks and cutlery handles.

Once the mass production of pianos had started, from c.1840 and up to the Second World War, Pratt, Reade & Co. became a huge importer of ivory, which it used in its business of manufacturing ivory topped piano keyboards for piano firms throughout America and elsewhere (it had a London office also). The company's original factory has been replaced by new housing stock (partly shown above) though the name Piano Works has been kept as a reminder of the town's long-gone industry.

With the growth of the piano trade, helped by the popularity of a new kind of piano, the player piano (or 'self-playing piano'), a new kind of town, Ivoryton, was set up in the area. But there soon appeared

Above: Ivory tusks on their way to the factory in the town of Deep River. It is estimated that in 1800, 26 million wild elephants roamed Africa. By 1930 it was 10 million, and today the WWF estimates the figure to be 1 million.

something unwanted: ivory dust. It was sometimes used as garden fertilizer, but it was said that when kids swam in the local ponds, they came out coated in the dust.

The ivory used in the factories wasn't always a uniform colour, nor was it particularly white. In consequence, huge greenhouses containing blocks of ivory were erected by the companies. Here the ivory was left to be bleached by the sun. On average, forty-five keyboards could be made from one tusk.

 You can tell these are ivory tops by the horizontal line just in front of the black notes (ivory key tops were made in two parts). As can be seen here, traditional ivory wasn't always ideal, it could chip or yellow and, over time, have a dip or hollow caused by the repetition of fingers depressing the key thousands of times over many years. Good quality synthetic tops, on the other hand, come in one piece, are easy to clean and wear very well. It was the Yamaha company who pioneered a new synthetic material – Ivorite – that came close to having all the best qualities of ivory, both in touch and with a 'grain effect' appearance.

INVENTOR (OF THE PIANO, BARTOLOMEO CRISTOFORI)

 Strangely, there is no statue to the widely accepted inventor of the piano (shown left). He didn't, in fact, even call his new instrument a piano, and in his lifetime he never came across an upright piano, a concert grand piano with an iron frame, nor a piano with the standard 88 notes.

A not-too-distant contemporary of JS Bach, the inventor of an instrument from which the modern piano evolved was an Italian musical instrument repairer baptized as Bortolomio Christofani (4 May 1655 – 27 January 1731). He was born in Padova (Padua in English), which is in northern Italy, close to Venice. By the age of 33, his expert instrumental craftsmanship had come to the attention of a Tuscan prince, whose former instrument restorer had passed away.

The Grand Prince of Ferdinando de' Medici had a great interest in mechanical things, notably clocks and musical instruments. Cristofori was employed to keep his collection of instruments, including the popular spinets and harpsichords, in good repair. The amiable Prince Ferdinando had around seventy-five musical instruments and was a great patron of music. A talented musician himself, he also knew or was in contact with leading musicians and composers of the day, including Scarlatti and Handel. Using his own wealth and often grand venues, he was able to put on annual concerts and operas. For his new employee, the relationship was mutually beneficial, for the Grand Prince gave Cristofori time, money and encouragement to experiment and develop his ideas and designs for various musical instruments.

In an effort to expand the range and capabilities of the Prince's collection, Cristofori first made two new smaller simple keyboard instruments for the Prince, before being more adventurous. The harpsichord at that time was useful for accompanying singers and other instruments but was thought to have its limitations. Notably, whereas most other instrumentalists had some control over dynamics – how soft or loud they played – with the harpsichord, the strings were merely plucked with individual inbuilt plectrums, the note sounding more or less the same no matter how hard or fast a key was depressed. There had been other earlier simplistic attempts to make piano-type keyboard instruments ('piano-forte' already had some usage), but it would be Cristofori who came up with the idea of having individual little wooden hammers (with the heads covered in deer skin) to strike the strings. We still have the evidence of his novel idea, as one of his pianos of 1720 can be seen in the Metropolitan Museum of Art in America (an image is shown below).

Although none of his earliest pianos have survived, evidence for the existence of his first piano can be dated to the year 1700 or earlier. Keyboard historian Stuart Pollens cites an inventory that describes his first piano. The bound inventory is dated 1700 and lists the instruments owned by the Serene Prince Ferdinando of Tuscany. Translated, it reads:

A large keyboard instrument by Bartolomeo Cristofori, of new invention, that produces soft and loud, with two sets of strings of unison pitch ... soundboard of cypress.

The inventory cited by Pollens goes on to mention the piano's dampers and hammers. Cristofori wasn't to know that his pianoforte, as it eventually became known, would kickstart a slow revolution in the world of music. He would not live long enough to ever see what, for centuries, truly became known as the piano though he did attempt to make more pianos in a similar style and when these (or copies) were sold and seen in Europe, in a small way it spread his name and invention idea. Similarly, renowned organ and harpsichord builder Gottfried

Silbermann (1683–1753) began to make fortepianos that closely followed Cristofori's designs; his first fortepiano was built in 1732, around the time of Cristofori's death. Composer JS Bach had criticised Silbermann's early attempts, saying the treble was weak and the touch unsatisfactory, but on a later visit he found an improvement in his more recent pianos.

Left: On display in the Metropolitan Museum of Art, Cristofori's 1720 54-note instrument – termed at the time as a Gravecembalo col piano e forte. A closeup is shown overleaf.

Two more of his early pianos are exhibited in Leipzig and Rome. There were harpsichords around at this time which were louder and more powerful instruments though did not really allow the performer to play in an especially expressive way, with degrees of soft and loud. In time, as the piano developed, people became very familiar with the piano's unique touch and tone.

The action in Cristofori's pianos overcame the difficulty of the hammer not bobbling/bouncing on the strings after the key had been depressed. Nor did the hammer stay on the strings ('blocking'), it was tripped just before reaching the strings, had enough momentum to continue and strike the strings, but was then held on a 'check' so that it didn't bounce back on to the strings. Held on this check, after the key had been released, the hammer would fall back to its original resting position ready to be played again. After Cristofori's first efforts, improved action designs were developed by various makers.

One piano that followed Cristofori's instruments, in the evolution of the piano, was the highly popular square piano, but they did not exist in his lifetime. The most universal type of piano in the world (and has been for over two centuries), is the upright piano, but Cristofori had been dead long before the first upright pianos came on to the market. That said, there existed in his lifetime the rather forgotten clavicytherium, a small upright harpsichord. Cristofori actually built one himself. London's Royal College of Music has such an instrument dating from *c.*1470 in its collection. The 41-note Clavicytherium reduces floor space and allows the musician to sit facing a vertical soundboard (whereas fortepianos have a horizontal one).

One can wonder why Cristofori's invention was so slow to take off. It was definitely a very slow burner, but of course its inventor may have seen it more as a development or experiment. He wasn't necessarily seeking to invent a brand-new instrument to introduce to the world, nor hoping for money-making rewards for patents and inventions. Additionally, the piano's predecessor, the harpsichord, had already been in existence for several centuries (there were also the smaller spinets and clavichords too). Harpsichords were Cristofori's bread and butter; behind and alongside him was Italian tradition, guilds and many talented craftsmen whose work and expertise were in demand. Moreover, the first pianos of Cristofori's time were not actually *that* dynamic when it comes to playing soft or loud, they were just different (in fact, the earliest pianos still had a hint of the harpsichord sound, despite having hammers). Because it had a different tone, there may have been musicians who liked it and others who were put off or subconsciously either suspicious or prejudiced against it. Neither was there any knowledge or tradition about piano technique, for the new instrument's touch would have felt strange to keyboard players, some finding it inviting and opening up new ideas, with others preferring to stick with what they were already comfortable with. In short, playing this new keyboard instrument required practice, experience and time. Cristofori's special idea of having the strings struck by small hammers led the way to the piano we know and love today, but his slightly more primitive first attempts were eventually superseded by the square piano (faster and cheaper to make and sell), and then the square's limitations – for it wasn't normally seen as a concert instrument – were overcome by the advent of both the grand and upright piano.

So history has seemed to overlook a possible second 'inventor of the piano', Johannes Zumpe, who came to London as one of the 'Twelve Apostles' – German keyboard makers fleeing the Saxony Seven Years' War (though the 'Apostles' term and number has its critics). Although not necessarily *the* inventor of the square piano, it was the London-made square pianos of Zumpe (which other makers copied) that really took off in Europe and America. Built from the mid-1760s, compact and much cheaper than most harpsichords, it sold in vast numbers for nearly a century (an example is shown on page 30). But the square evolved and came in different guises, from smaller, lightweight versions to more robust and sophisticated types later on. Certain later models even had a full-range keyboard similar to most modern pianos. They were highly popular, from royal families down to amateur musicians in private homes, but were eventually superseded by the grand and upright piano. The ubiquitous square piano, nonetheless, did not arrive on the scene until almost half a century after Cristofori's first pianos were made.

And then there was a 'third inventor of the piano': Hawkins. Many piano builders, both early on and later, were coming up with new ideas and styles for the piano, applying for patents (there even existed, for a time, a few 'double' instruments that could be played as both a harpsichord *and* a piano). Quite a few of these novel inventions and patents have been forgotten about as they did not survive in the trade, but we must not forget the person, considered by many, to be the inventor of the upright piano.

Somerset-born inventor, John Isaac Hawkins (1772 – 1855), would twice spend part of his life in America. He invented the mechanical pencil, also a system for sugar refining, and a water filtration system, but in 1800 he patented a small upright piano which had strings that descended down to floor level, an iron frame and suspended soundboard. Going by the name of 'Portable Grand', he sold one to Thomas Jefferson (perhaps inspired by Hawkins, it could be said that makers such as Southwell and Wornum did more to develop the upright piano).

With the arrival of the upright, many types would follow, often with confusing names. The cottage piano was the smallest (the Broadwood one, left, is unusual in having a wooden frame); cabinet pianos were taller (like cabinets), and others were marketed as upright-grands, presumably because they were iron framed and, although standing vertically, were believed to have all the superior qualities of a grand piano.

The first ever known piece of published music for the piano was written by Italian composer Ludovico Giustini and published in Florence a year after Cristofori's death: *12 Sonate da cimbalo di piano e forte detto volgarmente di martelletti.*

J

JAPAN; JOEL; JUILLIARD; JUNGLE; JOKES TOO

It might surprise the reader to learn that one of the world's most famous and expensive pianos, the Steinway, is actually made in Japan (also in China too). Okay, Steinway pianos are still made both in Hamburg, Germany and New York, USA, but it's true that two cheaper (starter?) options which carry the Steinway name, and can be purchased at many of the Steinway outlets throughout the world, are the Steinway Boston and Essex models. The Steinway-designed Boston piano is made by Kawai in Japan, and Steinway's Essex piano is manufactured by the Pearl River piano factory in China.

BILLY JOEL

The singer and pianist Billy Joel is the only non-classical artiste in the Steinway art collection. His portrait by artist and musician Paul Wyse was unveiled in Steinway Hall, West 57th Street in 2011. Joel was inspired to write his hit song, *Piano Man*, while working as a pianist in a piano bar in Los Angeles. Other portraits in the Steinway collection include those of Schubert, Beethoven, Wagner and Horowitz.

THE JUILLIARD SCHOOL

The famous Juilliard Music School (USA) possesses 260 Steinway pianos in practice rooms, teaching studios and performance halls. Among its long list of alumni is conductor Leonard Slatkin, also songwriter and performer Neil Sedaka, who won a scholarship there while a boy and went on to study classical piano at the Juilliard.

Right: The Alice Tully concert hall, part of the Juilliard School founded in 1905. It is an arts school, so in addition to being the former school of such diverse musicians as pianist Van Cliburn, singer-song writer Nina Simone and composer Richard Rogers, famous actors such as Robin Williams and, close friend, Christopher Reeve (of *Superman* fame) also attended.

JUNGLE

Pianos have found their way into jungles from time to time. Around the year 2000, eccentric but highly regarded explorer Colonel Blashford-Snell CBE, founder of the Scientific Exploration Society, was asked by a missionary priest if he could get hold of a grand piano for their church. The priest hadn't actually seen a grand piano for real (nor was he aware of how much they weighed) but had seen one in a magazine left by missionary workers. The colonel liked a challenge, for the priest and the Wai Wai tribe lived in huts in a very remote part of the Amazonian rainforest (in Masakemari – the place of the mosquito).

A 1930s Boyd grand was acquired from someone. It was taken to the capital, Georgetown, transferred onto a smaller plane which later landed on a grass landing strip. It then had to be manoeuvred by teak sledge and poles some eight miles through savannah and tropical jungle. The last bit was by a very large dugout canoe, which had to steer a way through rapids. It got there eventually and good use was made of it for a few years. Unsurprisingly, humidity and an onslaught of insects attacked the piano and eventually made it almost unplayable. A second trip some years after saw attempts to repair it, but also to supply an electric keyboard with a generator (though that, too, had its problems).

Below: Can he play? Colonel 'Blashers' at the keyboard (yes and no; he told the author he certainly had lessons as a child).

Older local tuners advised customers to stand their grand pianos in bowls of salt, to stop the insects crawling up the piano legs and into the piano (bowls of charcoal under the piano were also used to combat the humidity). Chappell's, in 1870, advertised special Oriental Pianos which were lined with perforated zinc to 'exclude insects'.

A much earlier jungle piano was an upright. Funded by the Paris Bach Society, in 1913 the French piano manufacturer, Pleyel, built the 'Jungle Piano' for organ scholar and polymath Albert Schweitzer. Built for tropical conditions, it was to be used in Lambaréné, Gabon, where Schweitzer's hospital had been established in French Equatorial Africa. Quite a few piano manufacturers claimed to build pianos especially suited to certain climates, but often they weren't that different to most ordinary pianos and could not really stand up to the more extreme climates and visits of termites seen in some countries. There is an interesting YouTube clip of Schweitzer playing his Pleyel piano which, all things considered, isn't unbearably out of tune (his poor cat sitting on the top seems to get short shrift).

Like many others, Schweitzer had concerns over the coming World War. The great polymath and Bach authority was tempted to turn away from his medical-missionary work and also his faithful instrument (which, unusually, came complete with organ pedals), but not for long. The following is taken from Joseph Gollumb's book *Albert Schweitzer: Genius in the Jungle*.

It all came closing in on him until the man who had always met his problems head-on found that for at least a breathing spell he simply had to turn away from it all.

There was only one escape for him that night. Schweitzer had not touched the piano that his friends of the Paris Bach Society had given him. It was not lack of time that had kept Schweitzer from it. If one must cut off a beloved past, it is best to make a clean cut of it, and Schweitzer had determined to let his fingers lose touch with music. Any reminder of their old skill would only bring pain without hope of relief. Now he felt that nothing of the past could bring such pain as did the present, and if music could help him forget it for the time being he was willing to pay the price. He went to the piano, raised the lid, sat down and played.

For the first time since creation the jungle about him heard Bach performed by a master, and if Bach's music, as Schweitzer has said, is an act of worship, then there must have been something of prayer in the playing.

Mostly unused, Schweitzer's first name was actually Ludwig. The hospital which carried his name was replaced with a modern one in 1981, but the 1952 Nobel Peace Prize winner's original jungle home was kept as a museum. Visitors to his former home reported his piano as still being usable in the 1960s.

Right: A modern made to order jungle-themed piano sold by Bath piano shop.

Q: I've got large ears, could you play a bit more pianissimo please?
A: Certainly. And what would you like me to play, *Nellie the Elephant*?

From Bach to Beethoven, Barton plays best for blind elephants! Since 2011, after moving to Thailand, classical pianist Paul Barton has regularly taken his piano into a Thai jungle elephant sanctuary and played for the blind, old or injured elephants. Apparently, when he starts to play, they wander over, a few sway gently and seem to enjoy the music.

JOKES (PIANO)

What do you call a laughing piano?
Answer: A Yama-hahahahaha.

Newspaper caption reads:
Caller: Good morning. I am here to tune your piano.
Lady of the house: My piano? I did not order a piano tuner.
Caller: No, but the gentleman across the way did.

Left: Bach at the piano.

Mozart as a child

Where do pianists go on vacation?
Answer: The Florida Keys

K

KEMBLE (AND OTHER 'GHOST' BRITISH MAKERS); KAPUT – PIANOS KNACKERED OR ABANDONED

The popular Kemble pianos were first built in 1911. A relatively late starter British piano manufacturer, yet they did well, lasting for nearly a century. In the 1970s they were producing about 5000 pianos a year (with many being exported abroad), but they finally closed in 2009 – perhaps receiving more media attention than they had ever enjoyed when trading. They were the last British company to make pianos on a large scale. Until 1968, they were in Carysfort Road, London (where the very successful Knight pianos were first made before moving out to Essex). After that, they relocated to Bletchley, Milton Keynes. A reminder of their north London origins is still there today, for close to the original factory is a road named Piano Lane.

KEMBLE & CO

Incorporating

B. SQUIRE & SON, LTD.
(EST. 1829)
SQUIRE & LONGSON, LTD.
ROGERS EUNGBLUT, LTD.

Manufacturers of the
MINX MINIATURE PIANO.

Pianoforte PRO PATRIÆ MUSICÆ *Manufacturers*

CARYSFORT ROAD · STOKE NEWINGTON
LONDON · N 16

Telephone & Grams
CLISSOLD 2446-2447.
A.B.C. CODE 5TH EDN.
CABLES:
KEMJAC. LONDON.

PARTNERS
M. KEMBLE
M.V. JACOBS

As with many piano firms around the world, whether in recent times or much earlier, it is often very confusing to know who actually built a given piano and where. Bechstein, for example, also make the Hoffmann piano – but this is done in a factory of theirs not in Germany but the Czech Republic; Zimmermann uprights – designed by Bechstein in Germany – are manufactured in China.

Returning to the Kemble firm, in 1933 they acquired Moore and Moore pianos. A similar company, Chappell's, had acquired Allison Pianos in 1929, and swiftly bought both Collard & Collard Pianos and John Strohmenger and Sons. After their move to Bletchley, Milton Keynes, with plenty of other British piano firms struggling with competition from Asian competitors, Kemble bought out and made the Chappell piano, but eventually Kemble was bought out by Yamaha (after 2009, the trusty Kemble was reborn in east Asia). Whether in the middle or late part of the last century, it is doubtful that customers buying a famous British brand of piano would have had any real understanding about which firm actually made their piano, regardless of the name it might have carried, as various piano firms made other well-known pianos under licence.

For the British piano trade, north London would prove to be a major centre for piano manufacture, notably in Camden Town and Kentish Town. This was partly due to the convenience of transporting timber and goods using the nearby rail, road and canal links. Between 1870 and 1914, Camden was said to be the centre of the world's manufacture of pianos, where many thousands were sent around the globe. At its height, the Camden area had in the region of two hundred piano manufacturers; often, local families would have two or more members in the trade.

Before the arrival of the large and well-established piano builders such as Brinsmead and Collard & Collard, piano building started on a smaller scale, often 'one-man bands' working alone or employing just a handful of workers. As the industry emerged and grew, it was quite common to see different piano firms working almost next door to each other – they weren't necessarily rivals because there was enough demand for new pianos for everyone to have a share in the market (Broadwood, for example, had their first premises next to the well-respected Kirkman piano firm). There was also surprising cross-pollination of ideas and experience with the founders of those firms which became household names. Be it British makers visiting Erard in France, Carl Bechstein travelling to England, or the Steinways leaving Germany and settling in America, they all learnt something by visiting and/or working with other firms. As an aside, it's also interesting that, with the exception of Ignaz Bösendorfer, all the founders of what became household names in the piano world must have been pretty fit and healthy men for those times. Founders Bechstein, Blüthner, Broadwood, Erard, Pleyel, and Steinway all lived beyond the age of seventy (as did the piano's inventor, Cristofori).

As the British piano trade grew, specialization came in, with peculiar sounding names for certain jobs. In addition to 'stringers', there were 'belly men' and 'silkers', for example, with the belly men making soundboards, and silkers (primarily women) working the silk used in the decorative fret-work seen on the front of some upright pianos (fret-work being another specialism, like making piano keys or hammers, where some firms did little else other than supply the piano trade with these parts).

We can see how some of these first men found their feet and saw new openings for themselves. The Knight company became a respected British manufacturer of upright pianos that were sent around the world. Its founder, Alfred Knight, started his career in more humble times. Born in south London's Camberwell Road, he was the son of a journeyman piano maker (also called Alfred). His great-great-grandfather had worked in the highly productive Broadwood workshop. When not in school, founder Alfred Knight helped out at the nearby Hicks piano factory in New Kent Road (South London, too, once had its own share of piano factories and dealers). Aged fourteen, Knight became apprenticed to Hicks in 1913. Upon completing his apprenticeship (five to six years then), he worked for Squire and Longson Pianos in Camberwell, which were later taken over by Kemble Pianos.

Alfred Knight (later awarded an OBE) set up his own piano firm in 1931 and moved it from east London to Loughton, Essex in 1955. It had all started with his ancestor working in the Broadwood factory (shown below), yet Broadwood's, the first piano manufacturer to hold a royal warrant, had started with one John Broadwood being

apprenticed to Soho, London harpsichord maker Burkat Shudi. Shudi had come over from Switzerland and started as an apprentice harpsichord maker for Tabel. He would go on to make harpsichords for Handel and royal families. Broadwood would later move his own firm forwards with pianos, many being used by very eminent people.

Whether it was Broadwood or Brinsmead, nearly all piano makers wanted to get their name 'out there', and the best way was through winning medals at exhibitions. There was often great rivalry among the makers, even more so with the 'foreigners' from abroad. It did turn slightly sour for some after the Great Exhibition of 1851. There were accusations of reporters showing an unfair bias to certain makers and countries.

With the birth of 'Piano Mania' during the Victorian period, all sorts of pianos were tried, anything to stand out or attract customers. Most of the new inventions have been forgotten about, among them were folding pianos, transposing ones and even a twin keyboard upright where two or more pianists could play the piano, back and front. But it is also forgotten that many entered the supposedly lucrative piano making or dealing trade and lasted only a short time, with some firms, large and small, becoming bankrupt; one such example is shown below. The image, right, is an 1871 painting by Walter Riddle, actually both a piano

tuner and talented artist. He married into the Moore family. The firm, started in 1837 in London's Bishopsgate Within, was taken over by Kemble in 1933.

The *London Gazette* of September 1842, announced under The Court for Relief of Insolvent Debtors:
William Binks of Leeds, out of business, previously Piano Forte Maker – In the Gaol of Rothwell.

A. Wilson, Peck & Co. of Sheffield (also Nottingham for a time) was founded *c*.1892. Partner John Peck had started his working life as a musician and piano tuner. By the 1960s the Beethoven House signage had been removed and it was no longer pianos that were the staple diet. Whilst they still had a piano workshop, school leavers could walk in and get jobs as trainee radio and television engineers.

Below: Street artwork in Peckham Road, South London. The building next to it, today named Piano Factory, was once owned by popular and thriving retailer W H Barnes (who didn't necessarily make the pianos which carried their name). They owned numerous outlets, including a branch in Oxford Street. As will be seen, there are other buildings which have a forgotten piano history.

Some former piano factories now make trendy bars and accommodation. The Old Piano Factory was once where Hopkinson pianos were made in London's Primrose Hill (it now has a road named Hopkinson's Place); it once employed 1043 men and trained 32 apprentices. The Chappell piano factory was nearby in Chalk Farm and is now used for apartments, the building incorporates some piano features in its interior design. The Piano Works building is in Piccadilly. Clapham's residential Welmar Mews is where former South London piano firm Welmar (*c*.1930s – 2003) used to be. British piano making definitely isn't extinct, however, for example award winning Cavendish pianos are hand-built in Yorkshire (and sell very well in Canada); while Cambridge piano firm, Edelweiss, make remarkably unique bespoke luxury pianos that are sold worldwide.

KAPUT – PIANOS KNACKERED OR ABANDONED

The images that follow (the first of which is of a former Chernobyl concert hall) show the sad demise of some once loved pianos, now destined for watery graves, attack or abandonment. In the very near future, the NSPCP will be launched to...

Left: A one-time fully-functioning grand used by high school students in Connecticut.

Above: One careful owner?

Below: An avant-garde version of Handel's *Water Music*? No, the sad-looking grand below mysteriously appeared in New York's East River.

STATISTICS
Vessel type: Player Piano with Sail auxiliary.
Builder: Gulbransen, 1925.
Launch Date: May, 2012.
Positive bouyancy, 400 lbs., approx., at launch.
Disposition: Sank, June 2012, in 30 feet of water, Lake Union Ship Canal.

L

LEGO; LYRE; LISZT; LADY DIANA

In 2020, Lego launched their latest addition: a grand piano (not quite a baby grand because it was for ages 6+). In reality, it doesn't actually play real notes, but can play music via a mobile phone. It can also be dismantled and the relevant piano parts have been faithfully reproduced (the piano contains a total of 3,600 pieces).

LYRE

The lyre is a small stringed instrument known for its use in Greek classical antiquity and later periods. In the piano world the 'lyre' refers to the pedal mechanism – called the pedal box by some – that is bolted on to the underside of a grand piano's key bed (which houses the actual keyboard). The name lyre came about because some grand pianos had the vogue of having the pedal mechanism resembling the actual lyre instrument (as in

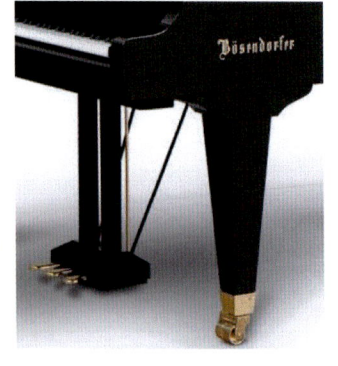

the first example above). The Steinway company also use a lyre as their company logo (for information on pedals, see page 133).

FRANZ LISZT (1811 – 1886)

The Hungarian composer, arranger, organist and virtuoso pianist Franz Liszt was born at just the right time for when the piano was beginning to go places. But of course, it was he who helped to push the evolution of the piano forwards from a small wooden-framed harpsichord with hammers (almost) to an instrument of ever-increasing power and efficiency suitable for solo performances.

The child prodigy Liszt was first taught by his father, Adam, when aged seven. His mother was Anna but it was mainly through his father that he got his early musical interests. Adam Liszt had been in the service of Prince Nikolaus II Esterházy, which allowed him to know Haydn and Beethoven.

Adam Liszt was a proficient pianist, violinist, cellist and guitarist, so it was natural that it would be he who gave his young son his first piano lessons. Within a year, Franz Liszt appeared in a concert where his young masterful talent was spotted by a group of wealthy people who immediately offered to pay for his education in Vienna. It was Czerny, a former pupil of Beethoven, who gave him his lessons there. He was fortunate in not being charged any fees by Czerny; in later life Liszt taught around 400 students across forty years as a teacher and mentor, but much to his peers' disapproval, never charged fees for lessons. In fact, in adult life he came to know and support the efforts of numerous musicians and composers who became household names, among them Chopin, Grieg and Richard Wagner (the latter would marry his daughter, Cosima).

It is unsurprising one biographer required three volumes to cover Liszt's life, as he led a full and active life which included many interests and contradictions. For example, he was a devout Catholic, one who dallied with becoming a priest, yet his religious or moral beliefs never interfered with his personal relationships – he is known to have had many affairs and fathered several illegitimate children. His daughter Cosima was from his relationship with Comtesse Marie d'Agoult, who left her husband and three children to live with Liszt; they would have a total of four children of their own. When this relationship eventually came to an end, he took his children and went to live with his mother (his father had succumbed to typhoid; Liszt, meanwhile, resumed various affairs after leaving the Comtesse).

It was after seeing the charismatic and virtuosic violinist Paganini around the year 1830 which inspired Liszt to stop performing in public and apply himself to rigorous practice so that he could aspire to be the 'Paganini of the Piano'. With his long and powerful fingers – despite some criticism of too much showmanship by some – he trail-blazed the use of the piano in solo performances (it is said the word 'recital' first gained coinage after his London concert in 1840). He is also credited with arranging for grand pianos to be positioned sideways-on, with the lid open to the audience, which allowed audience members to see his hands when playing – perhaps some were also intrigued by his rather well-known facial expressions (others claim Czech composer Dussek was the first to position pianos sideways-on).

Despite Liszt being a household name today and his image, in various forms, being seen across the media, we don't have any recordings of him actually playing. Some accounts describe Liszt's performances as being mesmeric and with awesome pianistic skills. For his pupils, he encouraged them to be individuals and didn't want them to play in his style. We do, however, have an account from 1831–32 when he was earning a living primarily as a teacher in Paris. Among his pupils was Valerie Boissier, whose mother, Caroline, kept a careful diary of the lessons:

M. Liszt's playing contains abandonment, a liberated feeling, but even when it becomes impetuous and energetic in his fortissimo, it is still without harshness and dryness. ... He draws from the piano tones that are purer, mellower, and stronger than anyone has been able to do; his touch has an indescribable charm. ... He is the enemy of affected, stilted, contorted expressions. Most of all, he wants truth in musical sentiment, and so he makes a psychological study of his emotions to convey them as they are. Thus, a strong expression is often followed by a sense of fatigue and dejection, a kind of coldness, because this is the way nature works.

Although rather infirm and with dropsy towards the end of his life, Liszt managed to get to England, where he played a number of concerts, including one for Queen Victoria at Windsor Castle. His life would come to an end, aged 74, after suffering from pneumonia. He is buried in the municipal cemetery in Bayreuth, Germany.

In 2001, the last piano owned by Liszt, an Erard grand, which he used in Italy for about fifteen years, was put on display at the Metropolitan Museum, New York. Paderewski had performed on it after Liszt's death, at a concert at the Vatican for Pope Pius X. The piano then disappeared until it was discovered again and found to be in reasonable condition. Measuring just short of seven feet and having eighty-five notes, Leslie Howard, pianist and president of the British Liszt Society said: *This Erard is the only instrument of Liszt's that today can be brought back to the concert hall and used for recordings and documentaries.*

Right: Liszt is seated at a Bösendorfer.

By artist unknown.

LADY DIANA

Lady Diana, or Princess Diana or Lady Diana, Princess of Wales, the People's Princess – or even 'Duch' to certain family members and friends – call her what you will, was actually quite musical and a competent pianist in spite of her modesty and shyness. Whilst Prince Charles (and future King) had been encouraged to learn the cello when young and later played it in the Trinity College orchestra, Diana had a much earlier and greater influence in the shape of a concert pianist in the family.

The family pianist was a servant of sorts, for Lady Diana's maternal grandmother was a Lady of the Bedchamber to Queen Elizabeth, the Queen Mother. Lady Ruth Fermoy (later Baroness Fermoy) studied the piano under pianist/conductor Alfred Cortot at the Paris Conservatoire. She cut her concert career short after marrying Maurice Roche (4th Baron Fermoy), though did occasionally give public performances. Baroness Fermoy founded the King's Lynn Festival, which has attracted many top musicians over the years and is in its seventy-second year.

After marrying Prince Charles, it wasn't unknown for Lady Diana to occasionally show her talents at the piano. While on a royal tour of Australia in 1988, Charles and Diana were visiting a Melbourne music school when Charles was encouraged to play a few notes on an available cello. Then a host turned to Diana and asked if she could play something on the piano. Although very sheepish, she surprised everyone by recalling and playing some bars from Rachmaninoff's second piano concerto.

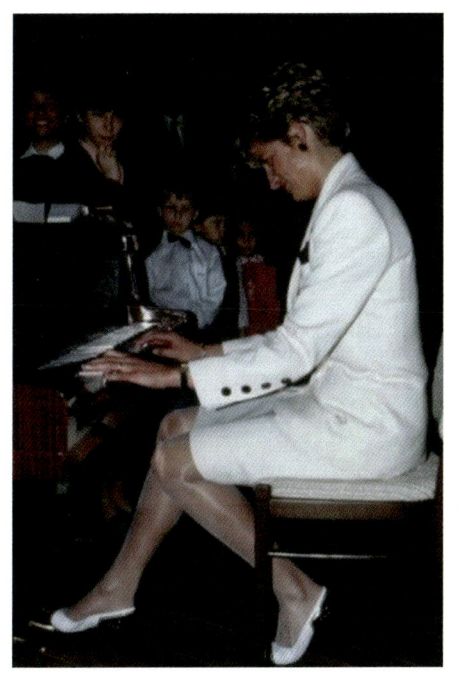

Lady Diana is shown in the image, left, playing for the children of an orphanage in Prague. *Daily Mail* royal photographer, Mark Stewart, explains how the photo came to be set up (in both senses of the term):

She would sometimes play tricks on us, so we decided to get our own back that day, we totally set her up. We knew the children were going to do a little concert for her, so we told them that Diana really loved playing the piano and that if they asked her, she'd come and play for them. So when they'd finished, one of the little children marched up to her and said, "We've just played for you, will you play for us?" She sat down and played Greensleeves absolutely beautifully for them. But as she left the room, if looks could kill! She knew exactly what we'd done.

The wife of Lady Diana's son, the Duchess of Cambridge, surprised many by also showing pianistic talent when she accompanied singer Tom Walker on the piano during her and Prince William's carol concert held in Westminster Abbey in 2021.

M

MIDDLE C; MYTHS; MISHAPS; MARKETING

If you hadn't noticed, Middle C on the piano isn't actually in the middle of the piano. The middle of the piano on a standard eighty-eight note keyboard is the space between notes E and F (however, at an earlier time, many pianos were made with a standard 85-notes range and on these pianos, the middle note *is* actually a C).

MYTHS

Luckily, some myths can be rather short-lived. In 1899, a report by the *British Medical Journal* stated that out of 1,000 girls who studied the piano before the age of 12, 600 were afflicted with nervous troubles in later life. Apparently, the study of the violin produced 'even more disastrous results' (the BMJ's report wasn't a lone voice at this time). Just in case some readers are under false impressions, below are various myths that have been around for a long while but still require dispelling.

Not all piano tuners are blind, in fact the majority aren't blind (many are registered blind but are actually partially sighted). Another linked myth is that being blind is an advantage for piano tuning or that it makes blind people better tuners. They weren't born with superior hearing, they have simply learnt to use this sense better; as with sighted tuners, the best tuners are normally those who have received a thorough and high-quality training. It isn't just good hearing that is required for tuners, they also need the skills that ensure the tuning will be stable for some time, not forgetting the many other requirements of a 'good' tuner, which include being: a skilful repairer, technician, being reliable and having good interpersonal skills.

No, the vast majority of piano tuners do not have perfect pitch, it is not needed when tuning pianos (even the very few with perfect pitch still need a tuning fork or electronic aid to check a piano's pitch accurately). Perfect pitch, of course, isn't completely perfect, and its definition and understanding are open to interpretation.

It should be noted that not all grand pianos are better than upright pianos. Baby grands, for example, have quite short strings and some may also have a more simplistic action, so a relatively tall upright from a good firm, using high grade materials, is likely to both perform and sound better than a small baby grand.

Pianists like to 'tickle the ivories' on their pianos. No modern piano today has ivory topped keys; the use of ivory, even in the best makes, was stopped from *c.*1960s onwards. Long before this, cheaper pianos never used ivory anyway, celluloid or another synthetic material was available and used long ago for mass-market pianos.

Milk is a good housewife's tip for cleaning piano keys and preserving their colour. No, milk wouldn't clean the keys, it might even make them sticky and unhygienic. A vinegar-water mix, on the other hand, used sparingly, can revive old casework. The covid pandemic, additionally, saw certain piano firms giving other 'expert' advice about cleaning keys, but some of it was unconvincing. My advice would be to take a damp cloth and apply just a little bit of common sense. Really, whatever you use has to be used sparingly and dried off. Simple things such as 'Mr Sheen' or a cloth dampened very slightly with soapy water is all that's needed (it's a piano, not an incubator, though piano pupils could be encouraged to wash their hands!). A useful way of quickly dusting a piano's keyboard, incidentally, is to use a large best-quality brush as used by a professional decorator (kept for use on the piano only).

<u>To be able to tune a piano you must be able to play one</u>! Not really and it does rather depend on your definition of 'play' (for when a cat walks up and down a keyboard, it too is playing the piano – though possibly Rossini's *Cat* duet accompaniment would be beyond it). It's rather like repairing a car, it's a mechanical process using tools, so being able to drive one is not absolutely necessary. That said, there are musicians who also tune pianos, some are keen amateurs, others semi-professional. Nearly all tuners can play chords, arpeggios and possibly a few tunes to try out the piano's tuning and performance but would probably answer that they know their limitations and 'don't play'.

<u>Moving a piano will make it go out of tune</u>. Carefully moving a piano within a home or other building such as a school is unlikely to make it go out of tune (unless it is placed against a new heat source, in direct sunlight or against an outside wall). It is the change of temperature and humidity when a piano is taken outside of a building and transported somewhere else that might make it go out of tune. It is worth buying a hygrometer, which are inexpensive, to place near the piano; its reading will show if the air is getting too dry or moist (ideal humidity range: 45 to 60%).

<u>All piano tuners are old and male</u>. Old piano tuners were young once and, for most of their careers must have been young or middle-aged, they are not all old but many enjoy a long career. In distant past decades, the occasional female piano tuner among the ranks existed and their number has increased in recent decades.

<u>An unused piano won't go out of tune</u>. It will, though it won't go out of tune as quickly as one that gets more use. Pianos go out of tune because of temperature and humidity changes; to a certain extent they can go out of tune for part of the year and then – almost – go back in tune again. However, the constant strain/pull on the tuning pins will slowly lead to the piano falling below concert pitch and its tuning deteriorating over time; on average, a piano should be tuned twice a year.

<u>Only pianos used for concerts are tuned to concert pitch</u>. Ideally, *all* pianos should be tuned to concert pitch because pianos and their strings have been designed to function best at concert pitch, which is known as A440 (440 Hz, the note A above middle C is vibrating at 440 cycles per second; once the A note has been tuned correctly to 440, all the other notes are adjusted to be in tune with this note). Concert pitch is really a standard pitch that was agreed upon around 1939 (and earlier in America). If the piano is tuned to this standard/concert pitch, it allows for

uniformity and enables the pianist to play along with recordings and other fixed pitch instruments (organs and xylophones for example). That said, if a piano has been allowed to fall a long way down in pitch or is very old, it may not always be possible to raise it to concert pitch. There are some awkward musicians and orchestras overseas, incidentally, who want their instruments tuned slightly higher (allegedly for a 'better, brighter tone'), but this does nothing to improve musicianship and consistency; when I contacted the Berlin Philharmonic Orchestra, for example, they told me that they tune not to A440 but slightly sharper, to A443.

<u>Cracks in the piano's frame or soundboard are serious defects</u>, if not attended to they could lead to the piano being written off. No, each piano and defect needs to be assessed on a case-by-case basis. Serious cracks may affect a piano's tuning stability or quality of sound, on the other hand there are many pianos with cracks and other defects but they do not affect the piano in anyway and don't require repairing (they might merely worry the piano's owner perhaps, though often the crack or other defect hasn't been noticed by the piano's owner until a tuner or technician has pointed it out – hopefully not in search of extra work).

<u>Acoustic pianos are expensive</u>. It depends on what one considers 'expensive' to mean. A brand-new piano's price may seem expensive, but as it has the potential to give over thirty years of daily pleasure and satisfaction (and even longer after reconditioning), can it be considered expensive? Most pianos easily outlive cars.

<u>Electronic tuning devices are better and more accurate for piano tuning</u>. High-quality tunings for millions of customers, top artists and recordings have been done by ear for several centuries by skilled aural piano tuners without the use of any electronic device. If one tuning was done with the aid of a tuning device, and the other by traditional aural tuning alone, it is highly unlikely that any musician would detect an improvement in the tuning (users who can only tune using an electronic aid tend to say, unsurprisingly, the devices are better and more accurate but this isn't necessarily true). For a good tuning, a sense of expert judgement and musicality is also needed, along with the expertise to correctly set the tuning pins (it is this skill that ensures the tuning will remain stable for a good length of time): all things a machine cannot do. One possible advantage of using an electronic tuner (done via a phone app or laptop and responding to screen information) is that it may be less mentally fatiguing or allows the user to work in noisier surroundings.

<u>The Steinway piano is the best piano in the world</u>. It must be so because, in their own way, they have said so and Steinways have been making pianos for over a century and a half. It's all a matter of opinion and interpretation of 'best', of course. Possibly it could be argued that Steinway pianos are the best for major concerts. Either by good fortune, design, an outstanding reputation or aggressive marketing, nearly all top concert venues throughout most of the world use Steinway pianos, so they appear to have a monopoly and, over the decades, if one concert venue has been known to use a Steinway, then others have tended to follow suit (a case of keeping up with the Joneses or Jacques?). The Rolls Royce car, too, has a high-class reputation and perhaps a certain amount of magic and history, but it doesn't

mean that it best suits the needs and tastes of all drivers who could afford to buy one if they wanted to. Because top artistes are seen performing on Steinways at many top events, it might appear that they are endorsing the instrument (and many do), but that doesn't mean they had much choice in the matter or that they would have chosen a Steinway at all costs (or that they use one at home). Lastly, when 'best' is used for describing a Steinway or any other piano, it depends on who is doing the describing. A piano tuner/technician may view a certain brand to be the best in terms of an instrument to work on and service (like cars, some pianos are pleasant to work on, others can be poorly designed or somehow rather awkward beasts). A pianist may judge a piano best by how it responds when being played (and certain repertoire may be better suited to certain makes or sizes of piano). The average age of a Steinway concert grand is normally under ten years of age, so older Steinways might not be deemed to be particularly good at all – it's all in the territory of taste and opinion, even motive. Someone selling a brand-new piano might well want to extoll its virtues, for example: they've never been made better than they are right now. Someone else selling an older reconditioned model might mention that the finest craftsmen and materials have been used on the instrument, which came from a golden period in the company's history (and the brand name on some pianos doesn't necessarily mean the piano was made by that company).

So 'best' can mean different things to different people. Yamaha, for example, reach a far wider cross section of the musical world, selling many of their instruments to private customers, schools and music academies. You will also see them in many top recording studios and on concert platforms. There are plenty of famous performers, additionally, who perform on Yamaha pianos and choose to have one for their own use at home. Yamaha also sell a wider range of instruments, so are definitely a 'best-selling' company. Steinway, along with numerous other well-known piano manufacturers, have had periods in their long history when the instrument proved to develop well-known (in the industry at least) faults. Bechstein grands of a certain vintage were known to develop cracks in the frame; in earlier times the Blüthner company unwittingly turned-out actions that used an inferior metal, requiring the action frame to be replaced. No matter what the piano, some manufacturers have produced the occasional star, while on other occasions producing pianos that were definite lemons and beyond any technician's help.

Yet there is a newer competitor on the market: the Italian firm Fazioli. They might not sell as many pianos as certain other makers, but the performance and quality of their pianos is like finding a secret holiday spot where you want to keep it to yourself and not spoil the pleasures it offers in case it becomes too well known.

Lastly, not all famous concert pianists own a Steinway or other very expensive grand. Pianist Sir Stephen Hough CBE, in his book *Rough Notes*, enlightens us:

The assumption is that we pianists will own the piano of our dreams, that we will have searched out the equivalent of a Stradivarius, found a generous sponsor or saved up to buy it, and then will spend happy hours playing rippling arpeggios up and down its pearl-white keys.

The truth is that most musicians I know have pretty rough pianos at home, not to mention the sound systems on which they listen to music (and balance their coffee cups). It's not so much a question of the cost of a great concert grand, although I found it hard to discover the current price of a nine-foot grand Steinway on the internet: 'If you have to ask, you can't afford', perhaps? It's more that I find it hard to work well on a gleaming young beast and I prefer to be hidden away in a back room somewhere with a gnarled, weather-beaten old Joanna. A concert grand is ... a concert piano; for me it feels too much as if I'm on stage performing. Practising is the workshop, not the showroom. Also, I don't want to own an instrument that makes every concert-hall experience a disappointment ... unless, of course, I can take it with me on the plane.

MISHAPS

The concert pianist is used to the minor mishap, perhaps an annoying mobile phone, squealing hearing aids or obnoxious cougher, but worse can happen. Like when Portuguese pianist Maria João Pires filled in for a concert in Amsterdam in 1999 at short notice. Pires sat waiting to start the Mozart concerto that had been scheduled for performance, only conductor Riccardo Chailly and orchestra struck up with a different (or 'wrong') Mozart concerto! Briefly sitting with her hands covering her panicked face, Pires then somehow dug deep and managed to find the 'right' concerto score in her head and came in just at the right time, going on to complete the whole concerto note perfect.

Sometimes the mishaps are scary in other ways, exploding light bulbs for example, or can be slightly more amusing, as in the case of pianist Alfred Brendel. Looking smart in his white shirt and tails at the end of a concert in Melbourne, it was slightly embarrassing for Brendel though did get the performer on to the first page of the *New York Times*. He rose to shake the hand of conductor Sir Malcolm Sargent, only to find he could not rise because his tails were inextricably attached to the piano stool. Despite people coming to his aid, eventually he had to take off the tails and do the customary handshake in his white shirt only.

The entertainer Sir Bruce Forsyth recalled a mishap when he was performing at the London Palladium early in his career. The expectant audience of 2,500, along with the waiting 30-piece orchestra, watched as Forsyth walked on to the stage to introduce the jazz pianist Erroll Garner. Forsyth became aware that the band leader wanted him to stretch out the introduction, he was mystified as to the reason why. He learnt later that the front piano leg on the grand had snapped while the stagehands had been trying to wheel it into position. Frantically, they had removed the broken leg and propped the piano up with a beer box hurriedly disguised with a piece of black velvet.

The above reminds me of a dog the pianist Horowitz once owned. Unofficially, it became known as the piano dog because it only had three legs. Well that's not quite true, but the unfortunate canine was accidentally run down by Wanda, Horowitz's wife, and landed up having one leg distinctly shorter than the others.

Jazz pianist Oscar Peterson was apparently a bit of a prankster from time to time, but as we learn here from an online review, so were some of his fellow musicians.

On one occasion, as the trombonist Bill Harris was about to play a ballad solo on 'But Beautiful' at a 1953 JATP concert, Brown had put a handful of small steel balls into the piano. These produced an impressive cacophony when Peterson tried to play and he had to reach over with one hand to try and pick the balls out of the instrument while accompanying Harris (badly) with the other. Harris, a giant of the trombone but a nervous player, was paradoxically a master joker. As he stepped back from the microphone he turned to Peterson and said, "One day. One day."

That day came on tour at the Rome opera house the following year. Peterson was due to sing a number with the trio. Harris had collected a tray full of glasses and empty bottles and put it on top of a ladder behind the back curtain of the stage. When Peterson began to sing 'Tenderly', Harris waited for the title word, pushed the ladder over and ran. The subsequent crash was satisfyingly cataclysmic. The stage sloped and so the bottles and glasses rolled down towards the footlights.

Elton John, having early success in the USA and recalling an incident at the Hollywood Bowl, wrote in his autobiography:

I had become completely obsessed with making a big entrance onstage myself, because it was the one time that I was really mobile, when I wasn't stuck behind the piano ... As I descended, the lids of five grand pianos sprung open, spelling out ELTON.

For the benefit of anyone who felt this was too subtle and understated, 400 white doves were meant to fly out of the grand pianos. I don't know whether they were asleep or too frightened to come out, but none appeared. As I jumped on top of my own piano, I found myself unexpectedly joined onstage by John Reid ... and a more sheepish-looking Bernie, running from one piano to the next, frantically grabbing doves and throwing them into the air.

Canadian concert pianist Angela Hewitt had finally got the piano of her dreams: a unique bespoke grand from master piano builders, Fazioli. She performed on her wonderful new piano at a recording studio, but afterwards it was unceremoniously dropped while being moved for its homeward journey. *The Guardian* reported:

A unique piano which was treasured by the Canadian virtuoso Angela Hewitt as her 'best friend' was broken beyond repair when it was dropped by specialist instrument movers. The expensive accident happened late last month after Hewitt finished recording Beethoven's piano variations at a studio in Berlin. She said it left her in such shock that it took her 10 days before she could announce the news to her followers. In a Facebook post Hewitt said her F278 Fazioli, the only one in the world fitted with four pedals, and worth at least £150,000, was 'kaputt'. She said: "I hope my piano will be happy in piano heaven."

The broken instrument was inspected by the firm's Italian founder, Paolo Fazioli, who declared it 'unsalvageable'. The piano's iron frame smashed when the 590kg instrument dropped as movers tried to lift it on to a trolley. The force of the break, compounded by the high tensions in the piano's strings, was so strong that it split the piano's lid in two. (12 February 2020; Hewitt now has a new Fazioli grand.)

Left: A road piano accident?

The pedals on the grand, left, no doubt failed to function well after the accident, but one piano tuner found himself squatting at the feet of Austrian pianist Friedrich Gulda in Buenos Aires and operating the sustaining pedal during a concert. The *Daily Telegraph* in 1954 reported that Gulda was playing Beethoven's Emperor Concerto when one of the pedals broke. It was a heroic tuner rather than emperor who apparently saved the day.

Left: Q: How was the concert? A: Smashing!

Apparently, it had rolled off the stage, but after being righted the piano was perfectly usable.

Something much worse happened to concert pianist Krystian Zimerman, who always took his Steinway with him to concerts. Following the 9/11 disaster, when he tried to export his piano to America for a concert, to security staff it apparently smelt of possible explosives and had to be destroyed.

Right: Who stole the show? A cat walks on to the stage at an Istanbul concert.

MARKETING

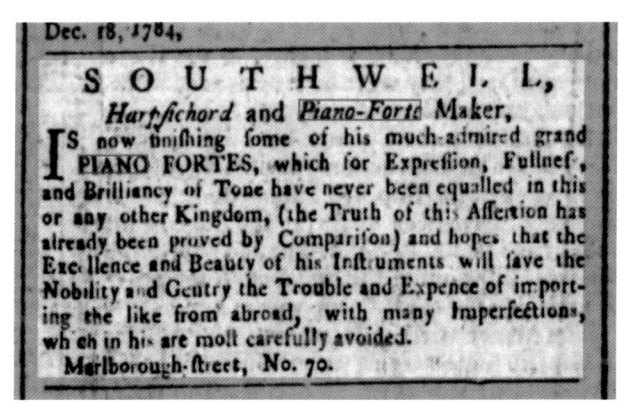

Dec. 18, 1784,

SOUTHWELL,
Harpsichord and *Piano-Forte* Maker,

IS now finishing some of his much-admired grand PIANO FORTES, which for Expression, Fullness, and Brilliancy of Tone have never been equalled in this or any other Kingdom, (the Truth of this Assertion has already been proved by Comparison) and hopes that the Excellence and Beauty of his Instruments will save the Nobility and Gentry the Trouble and Expence of importing the like from abroad, with many Imperfections, which in his are most carefully avoided.
Marlborough-street, No. 70.

Left: Pianos bought from abroad around the year 1784, apparently, were likely to have 'many imperfections' (maker Southwell, originally from Dublin, early on established a fine reputation). Half a century later, pianos were mass market commodities and could even be purchased on the popular 'Three Years' HP system.

Piano makers weren't slow in recognizing that if you have something to sell, you have to market it effectively. The early newspaper advertisement above shows two things: (1) the piano was evolving, it was a relatively new invention in 1784 that hadn't yet quite usurped the more traditional and faithful harpsichord; (2) initially, owners of pianofortes were wealthy people, it was mostly royalty, the aristocracy and gentry who acquired pianofortes as accessories, new technology, or pieces of art and furniture as much as for their musical potential. Cultured young females, of course, were encouraged to take up the pianoforte – possibly it helped to occupy their time, kept them home-bound to a certain extent but allowed their parents to show off their cultured and talented young ladies.

Times would move quickly. At first, for concerts and private use, there was a market for hiring pianofortes, or renting and then keeping the instrument, but as the boom in mass-produced pianos rose during the Victorian era, pianos in a range of styles and prices became much more accessible. Thus, piano advertisements focussed on quality, 'better tone', and size – there was a piano for every size of room (perhaps ever since with a class divide between owners of grands and owners of uprights?). Patents galore were taken out in a race for design improvements, which would see numerous disputes and court cases among the manufacturers.

A ROYAL PIANOFORTE.

A BARLESS GRAND FOR HER MAJESTY QUEEN ALEXANDRA.
John Broadwood & Sons, Ltd., cordially invite the critical, scientific and musical public to inspect the latest improvements in the Broadwood Pianoforte.
Catalogues of Artistic Pianofortes, in various styles, suited to any style of Drawing Room or Music Room, free on application from

JOHN BROADWOOD & SONS, Ltd.,
33, GREAT PULTENEY STREET (close to Piccadilly Circus), LONDON, W.

If it's good enough for the royals... But it also pays to name-drop as well – pianist, composer and conductor Dohnányi was well respected at the time. The invited audience who would have the pleasure of listening to a Broadwood, naturally, were people of the right sort: discerning and critical in a scientific and musical way.

Names of prominent pianists was always a worthwhile strategy, as was location and royal references. Thus, the Bechstein company early on had showrooms in Oxford Street and advertised as 'Pianoforte Manufacturer to the Queen'. Before the days of Steinway and others having their names emblazoned on the sides of their instruments, concert advertisements and programmes for such places as the Wigmore Hall included not only the name of performer but name of the piano being used (which, for the Wigmore Hall between the two world wars, was Bösendorfer, not Bechstein or Steinway). Marketing could be a tricky path to tread, nonetheless. While supplying more luxurious pianos to the rich and famous has always been one market, this alone would not keep firms in business. Thus, many chose to make and advertise pianos that had wider appeal, sometimes employing estate agent-type language where new pianos could be both cheap and of the finest quality.

A more recent pitfall was the lot of the Bechstein company, who made an impressive reproduction gold-leafed grand to commemorate their 160th anniversary. It drew negative comments in some quarters because it used genuine ivory on the keys (the company had wanted to be authentic and faithful to the original design, the ivory was legally sourced).

The 'angle' or something different was always needed even if there wasn't much of a difference or improvement really. Chappell's, in 1890, advertised a Yacht Pianino with folding keyboard – but how many per year were they likely to sell? Steinway, a year later, chose to use a scientific angle when advertising in *The Times*:

The supremacy of the Steinway Pianofortes – Artistic, Scientific, and Structural – in respect to all pianofortes in the world is a fact which has long since passed beyond the arena of discussion and is confirmed by every possible form of appropriate evidence. Scientific Pamphlets, containing illustrated lists, post free on application.

The advert, in layman's terms, is saying: if you want to be a somebody, get one of our catalogues and, hurry, buy one of our pianos!

Right: Class, culture, opulence, an instrument of courtship? Unusually for the time, it is a man seated at the piano. By 1893, Brinsmead produced nearly 2000 a year and employed several hundred workmen; their Wigmore Street showrooms displayed 350 pianos. Brinsmead said the premises were considered 'one of the sights of fashionable London'. One sell-hard Brinsmead idea was to book the Royal Albert Hall and have 50 grands being played by two pianists simultaneously (100 pairs of hands).

Right: Halifax Station, can you spot the advertising? Founded in Halifax *c.*1823, the Pohlmann firm were not directly linked to an earlier and possibly more famous London Pohlmann firm who made square pianos. Notice also the Thomas Rhodes & Sons firm above Pohlmann's signage. Although major suppliers of sewing and knitting machines, they too saw it worthwhile to deal in pianos and other musical instruments.

Dear Lady,

To you especially these pages ought to appeal, because you give the home the atmosphere of beauty and personality. Because to you the Piano signifies more than a mere musical instrument. You know that even if you don't play yourself, a Grand piano should be in your home for the sake of your guests, and you feel the artistic effect of its presence, which lends a charm to the entire home.

Left: From a Bechstein catalogue of 1928. A woman's place is in the home? And what kind of middle-class wife are you if you don't even possess a grand piano?

Right: Marketed on originality and size, the smaller print infers it can fit almost anywhere, any room any space.

STROHMENGER

The Outstanding Piano of To-day

Length only 4 ft. 2 in., with the tone of a 5 ft. Grand. Can be placed in any position in a room.

Cases specially designed to harmonise with various schemes of decoration

We shall be pleased to submit designs.

JOHN STROHMENGER & SONS, Ltd.

93-105, Goswell Road, LONDON, E.C.1 · · (Clerkenwell 2194)

Rushworth and Dreaper, the Liverpool organ and piano emporium founded in 1828, stated in a 1940s catalogue: *Most of us have experienced at some time or other the dullness of a social gathering which lacked a piano.* Their new pianos came with free delivery and a 20-year warranty.

<u>Pianolas:</u> Player pianos (self-playing pianos) are often called pianolas, though 'Pianola' was a trade name first used for player pianos made by the Aeolian company. In 1897, they acquired the rights from the New York-born inventor of the first practical pneumatic player piano, Edwin S Votey (who called his invention a pianola).

A player piano works pneumatically by having a perforated paper roll being driven either by an electric motor or by the operator pumping the bellows via foot pedals. The player piano allowed anyone to play music, you didn't have to be a pianist, and rose to its height in the 1920s. In its heyday, the manufacture of these instruments outnumbered the sales for conventional pianos.

They also created new jobs and helped to keep the piano industry afloat, but the advent of improved recordings and the radio hastened the player piano's decline.

Its early popularity saw many upright and grand player pianos produced by leading firms such as Bechstein. It was helped by the wide range of music rolls available and the fact that some of the rolls were recorded by well-known pianists and composers, including Rachmaninoff and Grieg (though as the roll of recorded music is being played on a completely different instrument and probably at a different tempo, it's doubtful a faithful reproduction of the recording is achieved). Numerous countries had music roll libraries where rolls could be exchanged, bought or loaned.

One well-known composer, George Gershwin, started to learn to play the piano when aged around ten and at the home of a friend who had a player piano. Author Stuart Jeffries tells us: *He slowly foot-pumped through a roll, and, placing his fingers over the keys as they were depressed by the roll-playing mechanism, learned the fingering for a piece. Two years later when the Gershwins had an upright piano, Ira recalled that brother George was quite the accomplished pianist: "I remember being particularly impressed by his left hand."*

Later, Gershwin used his keyboard skills to make and record piano rolls, earning him extra money while working as a pianist on Broadway. He made rolls quickly to capitalise on the popularity of tunes recently released as sheet music.

Votey's first player piano, it should be remembered, was a more simplistic mechanism pushed in front of any piano's keyboard; it was the development of a playing mechanism designed to fit inside any normal piano (as in the Victor image above) that led to the player piano's huge popularity.

The player piano/pianola, it has to be said, is not often favoured by piano tuners. Playing roll after roll, with the action getting quite a hammering, can wear the piano out much more quickly, particularly as a piano being played as a player piano can play more notes at one time than is humanly possible for any pianist. But the bellows and associated parts also make it awkward for the tuner to gain easy access to the tuning pins, strings and action.

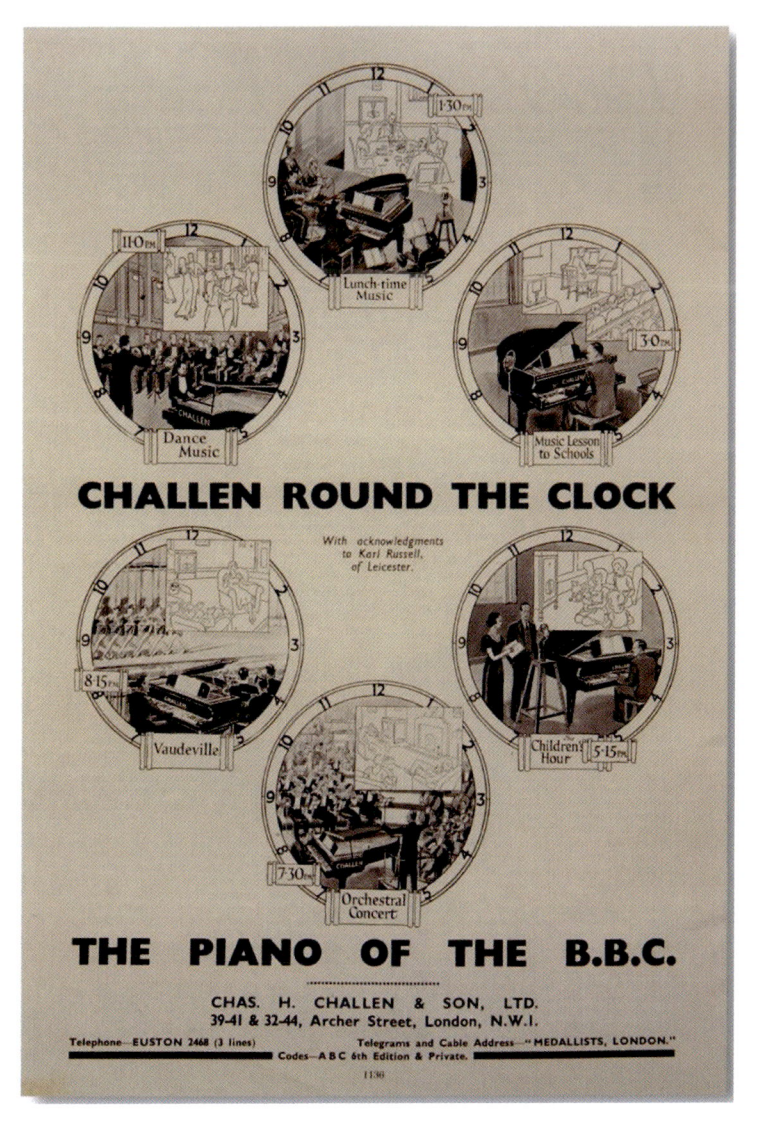

In the 1930s, British firm Challen did well to win a BBC contract, the BBC used over 180 of their pianos. The publicity also helped sales to the public though another marketing angle at this time was to build the world's largest grand piano, which had a length of nearly twelve feet. The BBC and certain other institutions later on became more open-minded and did not commit to one brand only. There are today, however, certain concert venues and music schools that, it could be said, are rather tunnel-visioned, believing that one brand only in their establishments gives their listeners and users the best experience possible. But what is the 'best'? If you deny pianists choice and dictate what they may use by imposing on them one brand only – musically, is this wise or fair?

Left: Perhaps shocking to those of a feminist persuasion?

Female models around a piano, as seen at a 1959 extension of the popular Ideal Home Exhibition held in Edinburgh.

SOME FOOD FOR THOUGHT

If you have music schools where all the pianos are of one make, it can be a double-edged sword. The students may have access to a good make of piano (so long as it is well maintained – though certain famous brands can be much more expensive to maintain), but it denies them access and experience of the contrasting tones and actions offered by other makers. It could make them less versatile and limit their taste and performance skills if tied to one brand of piano. (It is known, incidentally, that at least one music school presented itself as an 'All Steinway School', though the pianos were not from the Steinway factory, being the cheaper Boston and Essex models designed by Steinway but made elsewhere under licence for them.)

If a certain brand of piano is deemed to be the best piano, best for what? Best value, sound, best for reliability? Or best due to its known slow depreciation in value, structure, performance and tone quality? If a piano is going to be used in a college or other setting and for a range of musical genres by a wide range of pianists, a more reliable, versatile and robust piano may make another brand the best all-round piano to have rather than one originally considered to be the 'best'.

What you see pianos advertised for and what they actually sell for (or eventually sell for) are often very different prices. Similarly, selling a top-brand piano privately might demand a far lower price than when sold by a top-brand dealer. One 2019 published account of a Steinway-trained tuner/technician who, when working freelance, bought a second-hand Steinway upright for £1,650 is revealing. Completely overhauling it to showroom condition, he sold it to Steinway in London for £6000. It was sold in their showroom a few weeks later (having had nothing done to it) for just over £18,000. Yet currently and controversially, the term 'Stein-Was' has been coined due to the American company (it is said) saying that any Steinway not bought from one of their dealerships, or any older Steinway not rebuilt by them, cannot be called or recognized as a genuine Steinway piano.

There are many famous names quoted by piano makers as favouring their instruments, yet some of these names also appear on other makers' websites. In some cases, the artiste is not necessarily praising all of the piano maker's pianos, just the one they happened to have discovered and used (so not necessarily from the maker's recent stock). Glenn Gould almost worshipped a Steinway built in the 1940s, sadly it was dropped and he never used it again. He loved the 1895 Chickering grand piano he'd used in the family home and later had it moved to his apartment; later still, he chose to use a Yamaha piano for a number of recordings.

Even earlier, the great Paderewski, for a time, fell out with Steinway and started championing a rival, Weber pianos. In recent times, similarly, the *Washington Post* told of concert pianist Garrick Ohlsson being banned from using Steinway pianos after publicly praising a Bösendorfer he'd used in a concert. And on another occasion Paderewski, just mentioned, was due to play with the Chicago Symphony Orchestra, under conductor Theodore Thomas. The conductor saw it as entirely appropriate to support the local well-known piano manufacturer of the time by using one of their pianos. Knowing that Paderewski was considered to be a Steinway artiste at the time, somehow a Steinway grand was secretly placed on the concert platform just before the concert for him to use. Paderewski was unaware of the change of piano, but the conductor and orchestra received bad local press about what was considered to be a degree of disloyalty to local industry.

Most recently, concert pianist Angela Hewitt has made a point of preferring to bring her own Fazioli piano to her concerts. In other piano genres, too, some prominent pianists have shown a strong preference for the wide range of other high-quality pianos now available. Top jazz pianist Joey Calderazzo (a Blüthner artiste, as was Oscar Peterson), in an interview in the *Washington Post* in 2015, commented:

If I could have any piano I wanted, Steinway would probably be six or seven on the list. The problem is that each Steinway is so different, I have no idea what I'm getting. If you find a Steinway that's a good one, it's as good as any other piano out there. But one in 30 Steinways are good. And you have other piano brands that are actually kind of changing the game.

N

NORA; NEUROSCIENCE

Nora the piano cat, born in 2004, was a tabby cat from New Jersey. She was actually rescued as a kitten living on the streets (the animal rescue shelter was called, appropriately enough, Furrever Friends).

She had the necessary vaccines and did not succumb to any harmful viruses, instead a harmless video of her went viral after she was shown playing the piano on YouTube. *The Times* newspaper, in an online edition, reviewed her performance as being 'halfway between Philip Glass and free jazz.' It led to Nora's owner, Betsy Alexander, a music teacher, appearing on numerous talk shows, but while the appearances eventually petered out, Nora was said to still play daily duets with Betsy or one of her students. Although Nora was no fan of barks, it seemed she really did like Bach. Betsy Alexander, leaning more towards Beethoven, wrote a composition called *Fur Release: A Prelude for Paws and Hands*. (Dear Nora had a good innings and ascended to cat clouds and then heaven in February 2024.)

NEUROSCIENCE

It is well known that, for many piano owners who go out to work every day, the first thing they want to do when they reach home is simply close the door and play their beloved instruments. Some can't do anything else before this, as it relaxes them and allows them to 'switch off' (one London heart surgeon used to recommend his patients listen to Bach's piano works before surgery).

Recent research has shown that people with dementia and Parkinson's disease also benefit greatly from playing the piano as it helps to slow down symptoms and improve memory and brain function. As you will see, child pianists benefit also.

PubMed online report: Effects of music learning and piano practice on cognitive function, mood and quality of life in older adults.

Reading music and playing a musical instrument is a complex activity that comprises motor and multisensory (auditory, visual, and somatosensory) integration in a unique way. Music has also a well-known impact on the emotional state, while it can be a motivating activity. For those reasons, musical training has become a useful framework to study brain plasticity. Our aim was to study the specific effects of musical training vs. the effects of other leisure activities in elderly people. With that purpose we evaluated the impact of piano training on cognitive function, mood and quality of life (QOL) in older adults. A group of participants that received piano lessons and did daily training for 4 months (n = 13) was compared to an age-matched control group (n = 16) that participated in other types of leisure activities (physical exercise, computer lessons, painting lessons, among other). An exhaustive assessment that included neuropsychological tests as well as mood and QOL questionnaires was carried out before starting the piano program and immediately after finishing (4 months later) in the two groups. We found a significant improvement on the piano training group on the Stroop test that measures executive function, inhibitory control and divided attention. Furthermore, a trend indicating an enhancement of visual scanning and motor ability was also found (Trial Making Test part A). Finally, in our study piano lessons decreased depression, induced positive mood states, and improved the psychological and physical QOL of the elderly. Our results suggest that playing the piano and learning to read music can be a useful intervention in older adults to promote cognitive reserve (CR) and improve subjective well-being.

In an increasingly frenetic world, it is more important than ever to be able to focus. Playing the piano has been proven to help improve concentration, which helps in every area of life. According to psychologist Lutz Jäncke, learning to play a musical instrument has definite benefits and can increase IQ by seven points, in both children and adults. Furthermore, a landmark study of the impact of playing the piano before reaching seven years old was carried out by Gottfried Schlaug in 1995 and commented on in psychological papers. It was reported:

[The study] found that the 'corpus callosum' or the axons that connect both sides of the brain, was unusually thick in the child pianists. Schlaug's work had some sceptics to start with, and for one thing, people claimed that there might be a confusion between cause and effect. The corpus callosum might have been bigger initially in those subjects. However, further studies have given yet more insight.

Schlaug studied further, and while at Harvard Medical School was able to observe actual increases in brain capacity and the size of the corpus callosum among children who took up instruments. Specifically, two-handed instruments such as the piano were those that caused the most benefit. Having to play something different with both hands seems to sharpen the mind and build that brainpower.

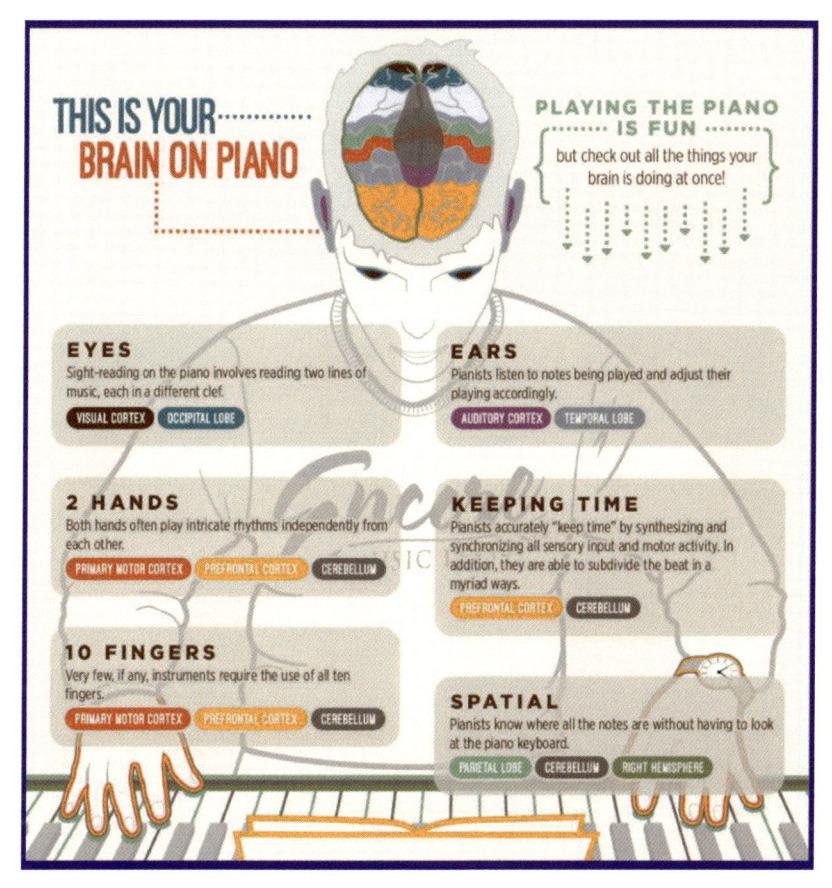

The rising Chinese-American freestyle skier Eileen Gu won silver and gold medals at the Winter Olympics in Beijing, 2022. She credited her latest success by the fact she had been playing the piano for nine years, which she enjoyed (and even took her keyboard with her to use between events). Reflecting on her long-term enjoyment of the piano, she felt it really helped her sense of rhythm with certain tricky manoeuvres.

From *Pianist* magazine: **Human Growth and Hormone** *Known as somatotropin, the human growth hormone is responsible for a range of functions including cell reproduction and regeneration. The hormone is essential for high-energy levels and cognitive functions such as memory. Luckily for the pianist, the act of playing the piano has been shown to increase levels of somatotropin!*

Life lessons *The process of learning the piano teaches some vital life lessons. No matter what age you are, the values of perseverance, patience, dedication and discipline are essential. Playing the piano nourishes these values and teaches you the importance of calm methodical practice.*

The results of very recent research at Bath University also reveal that a control group of adults without prior musical training benefited from having weekly piano lessons. Published in *Nature Scientific Reports* (November 2022), the study found that adults having piano lessons went beyond improvements in cognitive ability, it showed that participants also had reduced depression, anxiety and stress scores after the training compared to before it. The authors also suggested that music training could be beneficial for people with mental health difficulties.

OVERSTRUNG; OVERDAMPER; OSCAR PETERSON

All modern pianos, upright or grand are overstrung. Older and cheaper pianos, often pre-First World War, are straight strung (sometimes called 'vertically strung') and all the strings are parallel and run down the piano from top to bottom, all in the same direction. The overstrung upright piano in the image (the action has been removed), shows the thicker bass strings running diagonally from top left, partly across/over the steel strings. They finish close to the bottom of the piano, towards the centre and righthand side of the piano. This design allows for the bass strings to be both longer and nearer the centre of the soundboard, increasing the chances of an enhanced tone.

OVERDAMPER

Overdamper pianos are less common; they are older, cheaper uprights where the dampers (the felts that rest on the strings) hang from on top of a wooden rail on the action. In general, the dampers are not pushed on to the strings, as in modern uprights (where the dampers are both in front of and below the hammers), so tend to not damp as efficiently as the modern-day underdamper action with springs.

Left: A cheaply produced pre-First World War straight-strung upright. The treble strings, because they are short, are hidden, but notice how the bass strings on the left run down to the floor.

An overdamper action (right) can be identified by its bird cage appearance, with wires hanging down at the front (nearest the pianist; the diagram on page 7 shows the more common underdamper system used today).

OSCAR EMMANUEL PETERSON (1925 – 2007)

Born to parents who came to Canada from the West Indies, despite suffering certain illnesses throughout his eighty-two years, the jazz pianist Oscar Peterson rose through sheer talent to become an international star. Whilst his career eventually became synonymous with jazz and boogie-woogie, very early on he had begun his musicianship with classical training. When he much later took on certain jazz students of his own, he recommended they study the works of Bach, especially *The Well-Tempered Clavier* and the *Goldberg Variations*.

Peterson grew up in the predominantly poor and black neighbourhood of Little Bergundy in Montreal, where his young ears would have come across jazz sounds before his tiny hands had first touched a piano. Those same hands would later on comfortably stretch twelve notes, and his hearing go on to develop perfect pitch. But it was his dad who gave the five-year-old Peterson his first piano lessons. Dad played the trumpet too, and Peterson began to follow suit, but had to give up trumpet playing after being afflicted with TB. He was also helped in piano lessons by his sister (who became a local classical piano teacher). Perhaps it was her background and leaning towards classical music that led to her younger brother being conscientious in practising his scales and études.

A few years later, but still a young teen, Peterson received classical lessons from Hungarian-born pianist Paul de Marky, a student of István Thomáson, who had been a pupil of Liszt. Under his teacher's tutelage, he would happily practise four to six hours a day. An early break, however, came when he won a national music competition organised by the Canadian Broadcasting Corporation. Despite his father's encouragement to remain at high school (where he also played in a band), he opted to drop out. In time, Peterson became a professional pianist, starred in a weekly radio show and did numerous other gigs in hotels and music halls.

Building on his name and reputation, Peterson progressed to working alongside high-profile names such as Ella Fitzgerald, Louis Armstrong and Art Tatum (the latter he was always in awe of; upon hearing the blind Tatum for the first time, Peterson found he could not go near a piano for a couple of months).

From 1945 to 1949 Peterson worked in a trio and recorded for Victor Records (during his career he would release over 200 recordings). He would not be without occasional criticism, there were those who found his skills technically brilliant but perhaps lacking in originality. Count Basie, on the other hand, once said, "Oscar Peterson plays the best ivory box I've ever heard." And whilst he worked with a range of people – and could also sing and compose – his flexibility also saw him open to new ideas. He even recorded with Fred Astaire singing, and, in 1969, recorded with an orchestra two Beatles numbers: *Yesterday* and *Eleanor Rigby*.

He had a sad blow in 1956, however. When he heard his great friend Art Tatum was dying, he flew out to Los Angeles to be by his bedside, only to arrive too late. He was then given the news that his own father had also died that same day.

By now a Bösendorfer artiste, Peterson took part in many big concerts and tours. He'd suffered from some arthritis from an early age, also smoked both cigarettes and a pipe, and his large size tended to make mobility an issue. He had various children with a total of four wives. As age and health issues took a grip, things were made much worse after he suffered a stroke in 1993, which left him weakened on his left side. It took around two years before he was ready to play again, though not quite with the same skill he'd shown in his earlier career. Overseas tours continued, which saw concerts at the Barbican, and, in 2005, the Royal Albert Hall.

Oscar Peterson died of kidney failure at his Ontario home in 2007. He'd received countless Grammy and other awards and was even depicted on a postage stamp. A memorial sculpture of the pianist was unveiled in Ottawa in 2010 (shown overleaf).

The unveiling was by the late Queen Elizabeth II.

P

PERFECT PITCH; PITCH (CONCERT); PEDALS; PLACES; PREPARED PIANOS; PIANOLAS – see page 117

Perfect pitch, sometimes called 'absolute pitch', is the ability to recognize and name a note by merely hearing its frequency (or name all the notes in a chord), whether it is being sung, played on an instrument or coming from a non-musical source. Someone with perfect pitch needs no first note or reference point, they have all the notes (frequencies/pitches) in their head and so therefore can recognize and name them instantly, rather like most people can see and automatically name colours. If you are not colour blind, the moment you see something red, you don't have to think about it, you instantly know it's red. If you are out and about and hear a two-tone siren (once or sometimes still heard on trains and emergency vehicles), if a musician you would be able to name and identify the interval (distance of pitch between the two notes heard); someone with perfect pitch could do this but would also automatically know the correct name and pitch of each note heard.

The incidence of perfect pitch among the population is very rare, about 1 in 10,000 have it. The term 'perfect pitch' however, is open to confusion and interpretation; it has also, to a certain extent, evolved as ongoing research has brought new information and ideas to light. Moreover, the musical skills and the actual ability to recognize different pitches among those with perfect pitch can vary. In some cases it has been found that, as they get older, some have found that their accuracy has diminished. For those who have perfect pitch, it is not only musical notes they can identify, for many other sounds also have a discernible pitch (for example, you might find your hoover is unintentionally sounding a concert pitch C sharp!).

Examples where those with perfect pitch would relate certain sounds to musical pitches include: the ring on a bicycle bell, the chink of a glass being tapped, the squeal of car or bicycle brakes, or a whistling kettle. They may not sound particularly musical, but they emit a tone/sound that can be matched to a note on a musical scale (they have an identifiable pitch – pitch meaning how high or low a certain note or sound is, what particular frequency it is vibrating at).

The term perfect pitch is a slight misnomer because it is useful but not always perfect. It's near enough to recognise, for example, a concert pitch note of A 440 is an 'A', but if someone with perfect pitch had to tune a string precisely to A 440, they may not be one hundred per cent accurate, it may be very slightly up (sharp, possibly 442) or just under (flat, 439 or 438) when checked with an electronic tuner. This is why even the few piano tuners with perfect pitch still need to use a tuning fork or other device. And of course, perfect pitch can be a curse, as if an instrument is not tuned to concert pitch, when played most people wouldn't notice but to those with perfect pitch it would sound grating and out of tune (clashing with what are perceived to be the 'correct' notes in their head) even if the piano is perfectly in tune within itself. Others with perfect pitch have found it annoying when they know what key a piece is supposed to be played in but because the instrument in not correctly tuned to concert pitch, the piece sounds wrong, 'out of key' and somehow out of kilter even though the piano may be in tune within itself.

For those who don't have perfect pitch and need a starting note/reference point so they can work out the pitch of any other note being played, this is called relative pitch and is a skill any competent musician acquires. It is not the same as having perfect pitch, but the skill can sometimes be confused with it. This is in the same way that orchestral musicians who are used to regularly hearing the concert pitch A440 when instrumentalists are tuning up become very familiar with it, it stays in their memory if they are hearing it often. Others will recognize the opening note or chords of a favourite song and, having been told the names of the notes or chords, will be able to both recognize and name them, so they have a certain pitch memory but it falls well short of having perfect pitch (one theory, incidentally, suggests that possibly we are all born with perfect pitch but that for most of us the ability becomes contaminated in some way or we lose it in early life through not having it developed via regular listening and musical interaction. Those with perfect pitch have often had high amounts of musical interest and exposure at a very early age).

It has been noted that there is a higher prevalence of perfect pitch among speakers of tonal languages such as Chinese and Vietnamese. These languages have pitch variations for words, the same word can have a different meaning when expressed using a different pitch and intonation. Similarly, the prevalence of perfect pitch among people born blind from birth as a result of optic nerve damage (hypoplasia) is higher; it is also thought to be higher among people with an autistic condition.

For those with perfect pitch, there are advantages. For example, when tuning a guitar, they don't need to have a pitch source such as pitch pipes, they can just get on with it using the 'notes in their head'. They also recognize notes and chords as having certain characteristics; for some, they are familiar with family and friends having a familiar pitch range when speaking, their mother might mostly speak words within the range of the key of A, for example (the pitch range of a person's voice would pass most of us by). On the other hand, many every-day sounds have a recognizable pitch and it can be a nuisance: the hum of a fridge; a switched on hoover giving a continuous C sharp that won't go away; the howling wind that can be heard outside while trying to read a book is giving off an annoying pitch somewhere between G and G sharp. To most people it is 'just the wind', to those with perfect pitch, it has a name and frequency that gets in their head. Lastly, scientists believe even certain animals may have perfect pitch; bats and wolves, it is said, are able to make the most use of this ability for identifying mates and meals.

PITCH (CONCERT)

<u>Pitch: the short answer</u> – The musical pitch an instrumentalist or singer performs at relates to how high or low the notes are according to their frequencies. If a singer or string player is not playing with a fixed pitch instrument (such as a piano or organ, which can't alter their pitch and tuning while being used), their starting note can be any pitch they like. They can start on note F, for example, but it can be any 'F', as high or low a frequency as they like. The pitch they use can be more of a problem when instrumentalists want to play together (or play along to a recording), their F (and all the other notes) has to match and have the same frequencies as those of the other instrumentalists. Fixed pitch instruments can't alter their pitch during a performance, so for consistency they must be tuned to an agreed pitch which allows players to play together. The agreed pitch in this country and many others, although often referred to as concert pitch, is the internationally agreed standard pitch of A440 (used all the time, not just for concerts and obtained from a tuning fork or other device – more on tuning forks and also pitch on page 174). The note A above middle C vibrates at 440 cycles per second (also referred to as 440 Hz) and all other notes on the piano are adjusted to be in tune with the reference point of A440. Other instruments playing with the piano will tune to the piano's A.

<u>Pitch: a longer answer</u> – Musical pitch over history and even into this century has been something of a can of worms. The term concert pitch was used occasionally but it might have meant different things to different people (and in different countries). Interestingly, for a while 'concert pitch' was also quite a common phrase used outside the musical world. Similar to not being 'up to scratch', a person or object could be described as being on or off concert pitch (on form or below par).

Historically, musical pitch is linked to ego, competitiveness, stubbornness, politics, preference, finance and confusion. Generally, musical pitch rose during the nineteenth century. Composers such as Bach and Handel wrote music intended to be played at a lower pitch than our current standard of A440, about a semitone lower in fact (around A415 to A420 and sometimes referred to as Baroque pitch; some musicians playing on reproduction period instruments aim for authenticity by using this lower pitch).

It's partly a case that you can't please all of the people all of the time. While the pitch slowly rose (though throughout the nineteenth century there was quite wild inconsistency in this among opera companies, orchestras and organ builders), string players and singers began to find the rise in pitch detrimental both from a performance point of view and the long-term health of strings and vocal cords.

British Army bands had always used a much higher pitch than that used for pianos and organs; possibly it could be argued that the high pitch helped the instruments to perform at their best (it was said a higher pitch gave a brighter tone). However, it proved difficult – if not impossible – when organs, orchestras and bands were expected to play together. A letter in *The Times* of October 1884 highlighted the issue at Salisbury Cathedral. With 300 singers, organ and military band, it was stated that as the organ was tuned to the Society of Arts pitch, and the band to (the higher) Philharmonic pitch, it was almost impossible for them to play together.

There would be confusion in the terminology used and the understanding of what was the correct or most desirable pitch. In some instances, musicians and authors at the time unknowingly used the wrong name for a pitch, and for historians one has to be careful about which pitch they actually meant. For example, in the letter published in *The Times* there was reference to the Philharmonic pitch, but when the frequency of that pitch was altered later on, others began to use terms such as old Philharmonic pitch and new Philharmonic pitch. The old Philharmonic pitch was higher than the concert pitch used by most orchestras; it was close to British army military/Kneller Hall pitch, which was A452.4. Roughly between 1859 and 1885, musicians in most parts of Britain used what was called A435 French Normal Diapason pitch ('diapason' meaning tuning fork in French). This French/Continental pitch had been agreed on at an international conference held in Paris in 1859.

The inconsistency was fairly rife for organs, with well-known organs across the country being tuned to different pitches (some were also slow to move from a meantone temperament to equal temperament tuning). Behind some of this were the issues of tradition and money. If a certain organ has always been tuned to this pitch, why change it? And should the great Crystal Palace really be required to fall into line with St Paul's Cathedral or should they have the courtesy and common sense to match the Crystal Palace organ, played by many a famous organist (there was also the issue of time and cost to completely retune an organ's pipes)? Similarly, army band instrument makers Boosey and Company pointed out that the issue of pitch was not their fault, saying that the War Office had fixed the army pitch about twenty-five years ago. What wasn't always said openly was the possible inconvenience and financial implications if all band instruments had to conform to a new agreed lower pitch. In most cases the current instruments couldn't be tuned down, so would they all have to be replaced and new instruments manufactured?

Despite the need for some sort of consistency screaming out, for orchestras there too was an element of tradition and stubbornness, one orchestra (or their conductor) wanting to be unique or different to another; this could also apply to countries, where one country didn't want to be told what to do by another country. Yet even among piano makers within these shores, for over half a century or more, many of them had done their own thing, using the pitch they preferred for their own pianos. Maker John Broadwood, at least (or at last), galvanized action into getting British piano makers to conform and use a standard pitch for all pianos (piano tuners at the time often had to carry at least three different tuning forks so as to be prepared to tune the piano to the 'correct' pitch deemed by certain clients).

In July 1889, a meeting was arranged where all the influential names in piano manufacturing got together to agree on a standard pitch being adopted. A439 (nearly today's standard/concert pitch of A440) was put forward, and it was suggested that the high 'late Philharmonic pitch' (A454, equivalent to C540 and very close to the pitch used by British military bands) be regarded as the exception to the rule. Fortunately, all the leading manufacturers wrote to John Broadwood agreeing with his proposal, they included: Erard, Bechstein, Blüthner and Steinway. Only Mr Collard of that firm was slightly 'out of tune', writing to Broadwood and saying that his house had long tuned all pianofortes to their own 'medium' pitch, but that he would have a couple of his instruments lowered experimentally, and, if the rest of the trade decided, his firm would willingly come into line.

Partly linked to the increased broadcasting of music from different countries, 1939 would see the widespread adoption of a standard pitch of A440 (equivalent to the C above middle C being 523.3) being used – as it still is – in many countries. This agreement was the result of an international meeting held in London.

There would, of course, be practical problems regarding pitch, as there still is today. Pianos in cold environments tend to go sharp, whereas brass and wind instruments tend to play flat when the air is cold, and sharp when it is hot. Although we have greater conformity today, with A440 being a standard pitch widely used, there are the occasional overseas 'awkward customers' who choose to use a different concert pitch. The Berlin Philharmonic, for example, tune to A443. And the issue of what pitch should be deemed correct for singers and orchestras has never really gone away. In 1989, leading singers such as Plácido Domingo and Pavarotti added their names to a (politically motivated) Schiller Institute petition brought before the Italian government, asking them to lower the standard pitch at which all orchestras are tuned. The petition asked to lower the standard concert pitch of A440 to 432, claiming that 'the continual raising of pitch for orchestras provokes serious damage to singers, who are forced to adapt to different tunings from one concert hall or opera to the next.' The petition ended with a demand that 'The Ministries of Education, Arts and Culture, and Entertainment accept and adopt the normal standard pitch of A = 432 for all music institutions and opera houses, such that it becomes the official Italian standard pitch, and, very soon, the official standard pitch universally.' The request was, quite rightly, completely ignored.

In the examples below, the decimal point has mostly not been included as these are of little significance (age/condition and temperature also affects a fork's accuracy).

1751: A430, Handel's tuning fork (sometimes erroneously recorded as A422.5, it is actually a C fork of 512, which equates to A430 Hz, a low 'baroque' pitch).

1800: A455.4, Beethoven's tuning fork, a high pitch, nearly a semitone sharp of A440. (The sort of high pitch bagpipes are still inconveniently tuned to today.)

1820: A423, London Philharmonic Pitch; Westminster Abbey organ tuned to A422.

1826: Broadwood pitches: A433 low pitch, A445 medium pitch, A454 high pitch.

1834: Industrialist/amateur scientist, John Scheibler found that the average A on Viennese concert pianos was A440. At a conference in Stuttgart, the Congress of Physicists recommended the adoption of A440 (called Stuttgart pitch by some); this 1834 agreed pitch would turn out to be the widely used concert pitch of today.

1852: A452, average pitch of the Philharmonic Orchestra under Sir Michael Costa.

1853: A444, Paris Opera pitch; A 446, pitch used by Pleyel piano company.

1859: A435 (equivalent to C517), French Normal Diapason pitch set by the French government, having taken advice from experts and composers such as Berlioz and Rossini. Was widely used in England, also known as Paris/International/Continental pitch; it would later rise to A439 until today's A440 was adopted in 1939.

1878: New York Steinway pianos tuned to A457, London Steinways to A454. A454 was Philharmonic pitch, A452 British Army/Kneller Hall pitch. Both were labelled as old and new when their pitches were later changed. Society of Arts pitch was A449.

1880: A446 was Broadwood company pitch used for in-house tunings but not for public concerts (see date 1826; A446 is more or less their 'medium pitch' of 1826).

1884: The British Army/Kneller Hall (later becoming the Royal School of Military Music) pitch was high, at A452, and lasted until *c.*1928, before military bands began to use the standard pitch of A439 (French Normal Diapason pitch) used by most other musicians. (But there is evidence that many Salvation Army bands were still using the high military band pitch until the mid-1960s.)

1885: It was announced that French Normal Diapason pitch A439 (C522) would be used by Queen Victoria's private band and at state concerts. This new A439 pitch replaced the former French Diapason pitch of A435. It was said to be used by the Philharmonic Society (so was called New Philharmonic pitch by some), the Queen's-hall orchestra, and leading institutions in London. At a public meeting organized by the Royal Academy of Music it was agreed to adopt this new French pitch as a standard pitch for concerts in England. From 1899, leading British piano makers also agreed to tune their pianos to this new French Normal Diapason pitch of A439.

1919: It is said that in the Treaty of Versailles a concert pitch of A440 was adopted by all signatory nations, a 'concert pitch' was suggested but not specifically A440.

1939: A440 was the concert/standard pitch agreed on at an international conference held in London (used earlier in the USA and called American Federation of Musicians Universal Low Pitch). This standard was taken up by the International Organization for Standardization in 1955 (ISO 16, reaffirmed in 1975). The slight difference between this and the Diapason Normal was due to confusion over which temperature the French standard should be measured at. It was said that the older standard of A439Hz was superseded by A440Hz after complaints that 439Hz was difficult to reproduce in a laboratory owing to 439 being a prime number.

1949: The BBC's Third Programme, just before 6.00 pm, daily transmitted the pitch A440 (later, the dial tone on British telephone landlines would also be at A440).

PEDALS

Most modern pianos come with three pedals. The most used is the one on the right, the sustaining/damper pedal. The sustaining pedal lifts all the dampers off the strings at the same time. The left pedal is the soft pedal, also known as the una corda pedal on a grand. When this pedal is depressed, it moves the whole keyboard slightly to the right so that the hammers strike only two strings per note (or one string in the bass section) instead of three. Originally, early pianos often came with bichords – each note having two strings – so 'una corda' originates from the piano's early design where, when using una corda, the hammer really did strike just one string per note (today una corda is really due corde). On uprights, there is no real una corda. When the soft pedal is depressed, the resting position of all the hammers is moved forwards and slightly closer to the strings so that the 'blow' or travel distance towards the strings is shorter, thereby creating a softer sound. Additionally, on uprights the third (centre) pedal is known as a 'practice' or 'mute' pedal and gives a softer sound because the hammers strike a long strip of felt placed in front of the hammers before the hammers actually strike the strings.

For grands, the third (middle) pedal is called the sostenuto pedal and, in truth, doesn't get a great deal of use by most pianists. The sostenuto pedal is used to keep the dampers off the strings of notes (or even just one note) that are being played at that instant. The pianist is then free to play other notes while the notes just played with the sostenuto pedal depressed will continue to vibrate and have a sympathetic effect (because the dampers for those notes are being held off the strings). The pianist is free to play other notes (with or without the use of the sustaining pedal) and they will not be affected while the sostenuto pedal is depressed. Pianist Stephen Hough gives an example of an effective use of the sostenuto pedal in one of contemporary composer John Corigliano's études: *In the last moments of the Third Étude a minor third is held in the tenor register by this pedal, followed by the right and left hands scuttling up the keyboard in a non-legato, chattering flurry of thirds and fifths – a brilliant effect.*

PLACES (CONNECTED BY THE PIANO OR THE ACTUAL WORD)

Left: The Piano concert hall in New Zealand. It's not alone, there are other places around the world which are called Piano or have a piano connection. Hong Kong's Broadwood Road was named in honour of Lt. General R G Broadwood, Commander of British Troops in South China. He was a grandson of the founder of the John Broadwood Piano Co.

In the 1985 European Song Contest, Switzerland's 'Piano, piano', was voted twelfth place (because 'piano' is repeated 30 times?). They are on firmer ground with piano places, as Piano can be found in the Ticino region. Corsica, similarly, has a village called Piano, but unsurprisingly Italy has seven cities named Piano, also a village called Pianola (though 'piano' does also mean level, as in piano terra/ground floor).

Getting closer to New Zealand again, Australia has a riverside campsite called Piano Flat Campsite. Part of a conservation area that includes Piano Flat, the campsite gets its name from one-time colourful gold-miner Harry Selig. He was reputed to have been the first person to recover gold on the Flat and was also a piano-playing member of a local orchestra, formed to entertain its early settlers. Early on, the area was known as Piano Harry's Flat. (Additionally, there is a suburb in Queensland, Australia, called Brinsmead. It directly originates from the once well-known English piano firm, after one of the piano-making family – Horace George Brinsmead – settled there for part of his life. He pushed for sales of Brinsmead pianos but had also been a champion amateur boxer.)

Returning to Italy, with such places as Castel del Piano in Tuscany and even a Piano Lake (Lago di Piano) in Como, it also has a mountain range. One summit is named Mount Piano, while the higher neighbouring one is named Monte Piana (2,324m).

Steinway Pianos in New York has a street next to their factory called Steinway Street; this and a subway station were named after William Steinway (the station is shown overleaf). William Steinway (Wilhelm Steinweg as a child) started in the business as a soundboard maker but, a confirmed workaholic, rose to become a rich and powerful figure in America (his home 'Steinway Mansion' still stands). He was at one time chairman of the New York Rapid Transit Commission and had a say in the design of the subway system. He sadly died in 1896 of typhoid fever.

Seesen in Germany was the original home of the Steinweg/Steinway family and where they first started building pianos before settling in America. It still has a local park which was given to the community by the Steinway family. It is named Steinway Park and includes a popular 14.3-mile Steinway trail (shown below).

In England, London has a new Broadwood Road where the maker's Hackney factory used to be; there is also a Piano Lane – the Stoke Newington piano maker Kemble was once close by. Again, travelling north from London, there is a place called Roade, near Northampton, which has a piano link: Pianoforte Road. The road's name is a reminder of what was once there. From 1923, Pianoforte Supplies Ltd produced castings and fixtures for piano manufacturers; they also produced parts for the automotive trade. Finally, venturing on and into Scotland, there is 'Anderson's Piano' (Pass of Brander Stone Signals, Argyll & Bute). It is a place known to certain railway enthusiasts and is part of a unique signalling system.

Left: Hunai, China. A showroom designed to inspire the local population to take up the piano or any musical instrument. Right: The Piano café in Freshwater, Isle of Wight, the former home of Queen Victoria's official piano tuner (the front once served as a post office and piano shop).

PREPARED PIANOS

Prepared pianos are occasionally required by certain composers of avant-garde music. The instrument is used and therefore perceived in a different way, sometimes trying to get more unusual sounds and experiences out of it. One would hope that the best instruments aren't chosen when the performer is busy tipping grains of rice into it or actually plucking the strings with their sweaty fingers. Alternatively, as can be seen in the images left and below, assorted items have been placed on or between the strings in order to make them vibrate in an unfamiliar way.

Left: An unconventional tuning fork?

One well-known American composer of this unconventional type of music is the late John Cage, though his first attempt started back in 1938 when he'd been commissioned to write music for a dance. There was a grand piano at the venue but no room for a percussion group, thus the pianist was handed an assortment of items to use in the piano to get a variety of sounds. Interestingly, had Cage been around about a hundred years earlier, piano makers were quite adventurous. Many pianos came with not just two pedals, but several, or stops that allowed the performer to introduce different sound effects in the music, including bells, whistles or sounds imitative of other orchestral instruments (technology the nineteenth century way).

John Cage is possibly more well known for a piece in three short movements titled 4'33" (4 Minutes, 33 Seconds). The audience sits there in silence during the performance, yet so does the orchestra: absolutely nothing happens. Orchestra members sit and look at the conductor, who usually has a baton but also a timer on the lectern, no instruments are played however. The same piece was performed by pianist William Marx at the McCallum Theatre in California in 2010. Sat at the Steinway grand, he started the performance by closing the lid over the keyboard. In the breaks between the short movements, the lid was opened – the pianist having time to mop his brow before the next movement of silence. At the end, generally without much rapture, the audience politely clapped. Whether or not the piano was in tune, we shall never know. (There is also, incidentally, a real composition by Leroy Anderson for typewriter and orchestra that gets occasional performances.)

Q

QUOTES; QUEEN MARY'S DOLL'S HOUSE

I dream of music, but I can't write any because there are no pianos to be had here – in that respect it is a barbarous country.

Composer Chopin writing to his publisher Camille Pleyel while staying in Majorca.

****

We can only say that we shall be very happy to fall in with the request, and in subscribing 10s. towards such a praiseworthy object desire to state that all contributions shall be duly announced in these columns.

Monmouth Recorder of 1897, reporting on the Ladies Visiting Committee of the Swansea Workhouse, who had started a piano appeal for the 'poor imbeciles at the workhouse.'

****

I have been wanting for years to reform pianos, since they are as it were the very altar of homes, and a second hearth to people, and so hideous to behold mostly that with a fiery rosewood piece of ugliness it is hardly worthwhile to mend things, since one such blot would and does destroy a whole house full of beautiful things.

Artist Sir Edward Burne-Jones (an image of his own piano can be seen on page 68).

****

From the start there had been, as with Liszt and Paganini, serious over-reaction on the part of women to his playing, and it was reported with glee. Indeed the first pages of the newspapers were full of reports about frenetic girls going crazy at musical matinees. Soon Paddymania reached new heights, as three New York ladies embroidered musical phrases from his Minuet onto their stockings.

From an online article by Mikołaj Gliński about the charismatic pianist Paderewski. Streets had to be blocked off near his hotel; at his concerts, screaming female fans had to be kept at bay. Many wrote asking for a lock of his hair (quite easily done by his wife cutting a few strands from their dog and sending them off).

****

Why look for the spots on the sun?

Composer and virtuoso pianist Leopold Godowski commenting to a young man mentioning the wrong notes he had heard Josef Hofmann play in a recital. Godowski also gave his opinion after being invited to hear the offspring of a proud father play the piano. He wrote: *Your daughter is not without talent; she manages to play the simplest pieces with the greatest of difficulty.*

****

Too easy for children, too difficult for artistes.

Pianist Artur Schnabel commenting on the Mozart sonatas.

****

A fat little man like me, looking rather like a funeral director, and the piano has a little look like a coffin ... I have to hold them in attention ... I can't look at them, I can't make faces.

Pianist Arthur Rubinstein

When described as the greatest pianist of the century by one interviewer, Rubinstein explained that he got angry with such statements, saying only that there was no such thing as 'the best', only that some pianists (in some way) were different. The best artistes in any field are simply unique in some way. On commenting on his one-time friend but later on more of a foe, Horowitz, he wrote in his memoirs: *Horowitz returned to the concert life as the great virtuoso he always was, but in my view does not contribute anything to the art of music.*

****

The modern piano at its best is a super instrument; but alas! a great many have been manufactured at all times much less with a view to their excellence as musical instruments than to their imposing – and imposing is the right word – appearance as handsome articles of furniture. For, strangely enough there are many worthy folk utterly oblivious to the musical merit of the piano, folk who understand not a note of music...

From the *Pall Mall Gazette*, 1920

****

Bad music disturbs me, but wonderful music disturbs me even more.

Arturo Benedetti Michelangeli. Called 'a new Liszt' by Alfred Cortot, Michelangeli also said about performing in public (of which he didn't enjoy): *You see, so much applause, so much public. Then, in half an hour, you feel alone more than before.*

****

James Bond Buys a Piano!

The headline is the author's though a young Tam Connery (the future Sean Connery) did indeed buy an upright piano with his hard-earned cash. Among his numerous early jobs was as a French polisher. His father was against him using his savings on a motorbike so the Scot had a second-hand piano moved into the family's cramped flat.

****

Renzo Piano, the respected architect, has a musical name though no direct link with actual pianos. Designer of many impressive buildings, including the Pompidou Centre and Shard, he was also involved with the Parco Della Musica – when completed it was the largest symphonic concert hall in Europe – and is currently associated with the new Barbican Centre concert hall. As will be seen below, which is taken from an interview published by artforum.com, he had a close pianist friend.

That's why when you talk about building, you can't talk just about shapes or styles. It's also how you do it. You don't begin with some conceptual inspiration and then start thinking about nuts and bolts — you think about the two things together. And this is true for everybody. Maurizio Pollini, the great pianist, is a close friend of mine. Every time we talk, we talk about Steinways. You can't disconnect the music from the force of fingers on a keyboard. That is the power of technique: It is the link between idea and execution. Art needs that kind of depth, and so does architecture. That need will always connect them.

****

I detest audiences … I think they're a force of evil.

Pianist Glenn Gould. He also said: *My happiness is 250 days in a year in a studio.*

****

Named Andreas Ludwig Priwin at birth, the four times Oscar winner André Previn worked with the likes of Vaughan Williams, Olivier Messian, Oscar Peterson and Ravi Shankar, but would forever be known to countless television viewers as 'Andrew Preview' after his appearance on the Morecambe and Wise show in 1971. With Eric Morecambe attempting to play Grieg's piano concerto, some immortal lines unfolded:

Conductor André Previn: *You're playing* all *the wrong notes.*

'Pianist' Eric Morecambe: *I'm playing all the right notes – but* not *necessarily in the right order.*

The sketch's format had been tried out earlier, in the second series of *Two of a Kind* in 1963 (without Previn, Ernie Wise was the conductor). Eric Morecambe was said to be a passable pianist, he learnt to play the piano (and other instruments) at the insistence of his mother. The Grieg piano concerto had its snobbish critics who felt it was rather simplistic and not a great concerto, yet the masterful pianist and composer Rachmaninoff was a fan of it but considered it hard to master properly.

The versatile Previn, who had ten children and married and divorced five times (Mia Farrow being one of his wives), also had a great interest in jazz. In an interview in the *Guardian* (2019) about his work as a jazz pianist, styled after Art Tatum, he was asked if he considered jazz a parallel career to his classical work. He responded:

I remember reading an interview with Van Cliburn – who was indeed a very great pianist – and he said that he might try to play some jazz 'next year', and I thought, 'Oh really?' That infuriated me. I don't want to consider jazz something you do once in a while.

****

S for silence – Silence is the basis of music. We find it before, after, in, underneath and behind the sound. Some pieces emerge out of silence or lead back into it. But silence ought also to be the core of each concert. Remember the anagram: listen = silent.

A glance at the scope and wealth of piano literature makes us realise: this instrument works wonders. But the piano must be an instrument, not a fetish. It serves a purpose. Without the music, it's a piece of furniture with black and white teeth. A violin is, and stays, a violin. The piano is an object of transformation. It permits, if the pianist so desires, the suggestion of the singing voice, the timbres of other instruments, of the orchestra. It might even conjure up the rainbow or the spheres. This propensity for metamorphosis, this alchemy, is our supreme privilege.

Alfred Brendel KBE, from his book *A Pianist's A-Z: A Piano Lover's Reader*

****

And then there was Sviatoslav Richter. The first Richter recital I heard has stayed in my head as no other concert has. I remember everything he played and quite a lot of how he played it: an Olympian rendering of an early Schubert sonata; Franck's Prelude, Chorale and Fugue, in a performance so magisterial that I have never had the slightest inclination to hear anyone else play it; then, after the interval, an overwhelming account of the Liszt sonata.

This may all seem like nothing more than an exercise in 20-20 rose-tinted nostalgia. But golden ages really do exist. The 1960s and 70s came at the end of one such age. But those years were not some caprice of the pianistic gods. They were rooted in the European cultural history that immediately preceded them.

There can be no real dispute that the age of the pianistic 'lion' – the age of Liszt and Rachmaninov – is dead. It died with Vladimir Horowitz in the same month that the Berlin wall fell. It was the end of the era of the pianist as star, an era in which pianists could be seen as demons possessed by brilliant and magical technical skills.

John Kettle writing for the *Guardian* in 2002

****

You play the organ. What does playing this instrument teach us about playing the piano?

Frankly, the organ and piano have nothing in common! A cello is more closely related to the piano. It has a sonorous sound that has been described as being like an extension of the human voice. To my mind, the piano is not a percussive instrument.

However, based on experience, it is my belief that a well-trained and developed technique as a pianist is essential to being a superb organist. Liszt and Schumann composed outstanding pieces for the organ that cannot be performed without a well-developed piano technique. Remember, the organ itself is not only an instrument of the Baroque era; it's also the great Romantic instrument of Franck, Widor and Vierne.

You teach the piano as well. What do you think makes a good teacher?

That's easy: it is to draw out the personality of each student and to treat each in an individual way. Often, when I am teaching the same piece, I will say and recommend totally different things to my students so they are able to find their own voice – one that is authentic and unique to them.

Concert pianist Burkard Schliessmann interviewed in *Pianist Magazine*

****

The purpose of the inquest hearing was the coroner giving fair warning that if you have a claim then come forward because he will be making a decision at the next inquest. [Published 2017 in Shropshire]

Statement regarding a haul of gold sovereigns found hidden in an old piano by a piano tuner. Later said to be worth £500,000, they were officially declared 'treasure'.

****

Alison Wheeler, 39, was dismissed after officers made allegations that she failed to use her CS spray or support fellow officer Rory Channon during a struggle outside a police station in Walton.

Ms Wheeler, a former opera singer, claims that Surrey Police is a 'sexist and ageist' force and is seeking £350,000 in compensation as she says she did intervene.

Ms Wheeler denies accusations that she did not get involved when colleague PC Channon was assaulted while attempting to arrest one of up to five men who were fighting outside the police station. It is also alleged that she failed to use a radio to call for backup.

She said CCTV footage shows she called for backup and did get involved in the fray, actions which she claims were appropriate and that her dismissal was a result of superiors thinking her 'too posh'.

Ms Wheeler claims she had been the subject of bullying as she owned a grand piano and had been an opera singer, adding that her bosses 'lacked courage, and failed to take appropriate action to support her as a colleague'.

From *Police Professional* newspaper, June 2009.

****

The BBC is trying to kill its jazz listeners.

October 2011. Miles Kingston writing in the *Independent* about the BBC's cavalier and disrespectful treatment of its jazz audience, citing (among other things) the incorrect records being used and some even being played at the wrong speed.

****

It is a room in which I have experienced many conflicting emotions, from the tremble of nervous anticipation as I warm up on the well-worn piano to the exhilaration and relief at the end of a concert, a glass of chilled champagne in hand: 'Darling, what a performance!' 'I've never heard that piece played like that!' 'How do you think it went?' 'Dear, I'm speechless!' ... some of the classic phrases expressing dissatisfaction or bitchy venom while still appearing to offer a crumb of praise to the hungry, self-doubting artiste.

Reflections on concerts and the Green Room (used by artistes before and after concerts) in the Wigmore Hall, taken from *Rough Notes* by Stephen Hough CBE.

****

It is not known if the piano is for Boris or Carrie – but the Prime Minister did have one in the marital home he shared with his second wife Marina in Islington until they split last year. When he was Mayor of London he said his biggest regret in life is failing to make it as a famous rock musician. He said: "I think I regret it bitterly. I tried at school to master the guitar with a view to becoming famous and it was hopeless. And I thought, right well I'll master the piano and that went even worse."

The *Daily Mail* on the then Prime Minister's piano being moved into Downing Street.

****

And when my dad wasn't around, I played Little Richard and Jerry Lee Lewis songs on the piano. They were my real idols. It wasn't just their style of playing, although that was fabulous: they played with such aggression, like they were assaulting the keyboard. It was the way they stood up while they played, the way they kicked the stool and jumped on the piano. They made playing the piano seem as visually exciting and sexy and outrageous as playing the guitar or being a vocalist. I'd never realized it could be any of those things before.

From Elton John's autobiography *Me*, 2019

****

Piano House is a type of house in DayZ Standalone. This structure takes its name from a blood-stained grand piano that can be found on the first floor. This building spawns a variety of civilian-tier loot, including Food and Drink, Weapons, and Equipment.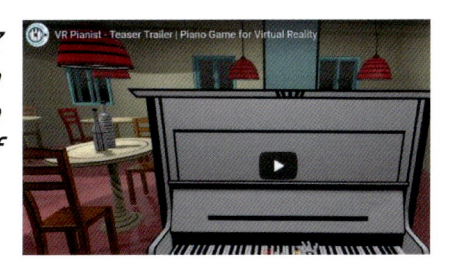

Pianos can now be found in the virtual world too. Another example is a fantasy piano in the PlayStation game *Shinra Mansion Piano!* or you can play a virtual piano using hands or controls at vrpianist.com.

****

"As soon as I walked into the lobby of one hotel," he said, "I immediately realised something was missing – but I couldn't put my finger on it."

Colin Bennett, a manager of Starwood Hotel Group, speaking to journalists. Three people had strolled into the hotel dressed in overalls and wheeled the grand piano out and down the street, they or it never to be seen again.

****

Then on the musical side I've been picking up the piano as well. I find it a very good way to put all my emotions in one place and it's a great outlet for me because, day to day, I have to put on a brave face a lot of the time. I think it's just a great way to release everything.

Tennis player Emma Raducanu in a Sky Sports interview, August 2024. A contrast with French philosopher Voltaire, who preferred the harpsichord to the newcomer, the pianoforte, describing it as *nothing much more than a tin kettle.*

QUEEN MARY'S DOLL'S HOUSE

Designed by Sir Edwin Lutyens and built in the early 1920s, the Doll's House was built for Mary, the wife of George V. Astonishing detail can be seen, it also has electricity, a working lift and plumbed-in water for the bath and toilet. Many of the fixtures and fittings were designed or made by the leading famous companies of the time, hence the grand piano was made by the Broadwood company (who were first appointed a royal warrant holder by George II – the grand actually has ivory and ebony keys). Well-known authors Rudyard Kipling and Sir Arthur Conan Doyle (owner of a Broadwood himself) wrote stories for the miniature books in the library. Similarly, composers Holst and Delius contributed miniature works, though apparently Sir Edward Elgar refused to contribute anything. The Doll's House was put on display at the British Empire Exhibition, 1924 – 1925, to raise funds for the Queen's charities.

Designed at a scale of 1:12 (one inch to one foot), the Doll's House is just over 3 feet tall and can be seen at Windsor Castle. Its piano has real strings and, with light fingers, can be played.

Below: Yes, there was even a children's Broadwood upright in the house.

THE QUEEN'S DOLLS' HOUSE   Grand Piano in Drawing Room

R

REGULATING; REMOVALS; ROYALTY

'Regulating' (or 'regulation') refers to the process of using special tools to adjust the piano's action so that it works at its best: evenly and efficiently throughout. The fact that on a piano all the keys appear to look perfectly level, for example, doesn't just happen, they have been set up that way (and over time may not stay that way). On a piano, some notes and sections get played more than others, so the keys that are played the most will have more wear and tear, which might result in looseness and noise when certain keys are played, so adjustment (regulating) may be required. For example, the notes that are played the most will have felts that have got thinner and springs that have got weaker; for evenness of touch, they will require some regulation or replacement. Other adjustments might include ensuring all the keys have the same weight and depth of touch ('key dip') and that the dampers all lift off the strings at exactly the same time when the sustaining pedal is used. Moreover, the precise moment a damper begins to lift off the strings can also be adjusted to enhance touch and performance. A tuner/technician might do some ongoing regulation on some notes during a normal visit, perhaps easing sluggish notes or tightening keys that are too loose. But a tuner/technician can also spend a longer time and adjust the entire action to the preferences of individual pianists, ensuring it has good repetition, or making the touch either heavier or lighter.

REMOVALS

Despite the heavy unwieldiness of them (for Laurel and Hardy in the 1932 film *The Music Box* but also for all of us), the earliest piano owners and firms always found a way to transport their instruments. Newspaper adverts of pianofortes for sale show numerous places of the empire, Calcutta for example, having pianos for sale soon after they were being manufactured. *Treasure Island* author, Robert Louis Stevenson, loved his piano and in 1891 had it sent out from Scotland to his new home in Samoa, where he played it every day. The Polish composer and pianist Frédéric Chopin ordered his Pleyel piano from the French maker and had it delivered to his quarters in Majorca. There were problems with bureaucratic delays and having to pay import taxes. The piano travelled in winter but it got there in the end, having travelled through heavy rain and over seemingly impassable tracks with a stubborn mule preferring to sit and wait on its haunches, yet it arrived in very reasonable condition (the composer could never have predicted Warsaw's main airport would be renamed Chopin Airport in honour of him around a century and a half after his death). Elsewhere, whilst servants in India would have been roped in to assist the camels transporting pianos to summer cottages in the Himalayas, later on and in other places of the empire and beyond, a 1915 guidebook tells us about how well-off Bolivian residents in Sucre deployed four mules to transport upright pianos over the giddy heights of the Andes.

People have often attempted moving pianos themselves, such moves can go with success but at other times accidents happen. Very occasionally, sadly, things can even go wrong for reputable piano removers who use all the correct gear (for example, webbing, dollies and shoes – a shoe/skid board is a wooden frame used to protect and house a grand turned on its side for removal; dollies, called bogies by some, are the specialist trollies).

In 2007, a Devonian music festival made a rather undeserved name for itself after reaching the national media, it became known as the Dropped Bösendorfer Festival. Founded in 2001, the Two Moors Festival Spring Concert Series was started partly in a bid to boost tourism and morale after the Foot and Mouth outbreak; it proved so popular that it became an annual event. They had started by hiring a grand piano for each concert but by 2007 had managed to raise sufficient funds to purchase a nine-foot Bösendorfer concert grand being auctioned in London. They were fortunate in that the auction happened to coincide with a larger auction in Europe, which drew prospective bidders away and helped to keep the asking price of the piano down. That said, the piano had still cost them £45,000.

It was taken from the auction house in London by lorry down to South Molton. The procedure for getting the large grand onto the tail-lift at the rear of the lorry so that it could be lowered to the ground, proved tricky and the piano had to be angled and 'jiggled' slightly to get it on the tail-lift. Things happened very quickly. It seems that part of the rear end of the piano had not quite cleared the lorry's rear framework so that as the tail-lift began to descend to the ground, part of the piano clipped part of the lorry's framework and quickly toppled off the tail-lift. It bounced on to the sloping drive. It kept going and because there was a bank with steps, it flipped right over and landed on its lid. As one of the removers recalled, "It gave a death rattle then there was a deafening noise, one hell of a crash and all its notes went at once. It fell about thirteen feet in all."

One of the concert organisers had been there with a camera to record the 'happy event' of the delivery, but the photos would be used instead for insurance purposes. The grand was rather ignominiously righted by the use of a farmer's front-loader tractor and returned to London where its damage could be assessed. The festival organizers managed to get a replacement grand in time for the festival. They were later on even luckier in getting a second greatly reduced concert grand from the Bösendorfer firm in Vienna, who arranged to have it delivered ready for the next festival.

With the Duchess of Wessex as its patron, the festival is still going strong and they are very proud of owning their impressive replacement concert Bösendorfer.

Whereas the average upright piano weighs around 140kg (308.647 lbs), a concert grand can weigh 544kg (1200 lbs, around half a tonne).

Above: The prized instrument is unloaded after its long journey.

Right: It was all going so well...

On this and the next few pages are a variety of removal scenes from across the years, though the one below is a very early photo from the Viennese Bösendorfer firm. Notice the removed lyre and three upturned piano legs.

People moving pianos rarely arouses suspicion, yet crime can occur. A case was reported where a pianist had not recently had a new German grand delivered from that country. He was in the process of moving home and the piano hadn't been moved to the new address. He arrived at his old home to find the piano missing. The new 'owners' of the grand happened to see some sales paperwork wedged in the piano, prompting them to contact the owner via a useful website they found. The illegal removers had sold it on quickly to a piano warehouse, who had asked few of the right questions before reselling it. Fortunately, the original owner was reunited with his piano. The site (which covers all stolen or missing instruments) is: www.musicalchairs.info

Right: French removal men at work. Firms then and now manage to push, pull and squeeze pianos into all sorts of spaces. In recent times, a Blüthner grand was moved by G&R Removals from a London showroom to the 58th floor of the iconic Shard building (appropriately enough, designed by Renzo Piano; possibly the highest piano in Europe?).

Accidents regarding pianos are much rarer than they once were though people often underestimate how easily an upright piano can lose balance. Back in 1894, the *South West Daily Post* reported on a school cleaner in Nailsworth who was working away. One of her children was playing nearby as she tried to move the piano. The castors got caught in the wooden flooring and it toppled over, landing on both her and her young son, who sadly died from fatal injuries (*Killed by Piano* ran the stark headline).

Left: A back-breaking business? In 2019, extreme adventurer and fitness coach Max Glover had already run a marathon with a car strapped to his back, so running up a mountainside the following year with a Kemble upright on his back was no problem, surely? Despite some earlier injuries and illness, he succeeded in carrying the piano up Garth Mountain in Wales to raise funds for the Royal Brompton and Harefield Hospitals Charity. The arduous 2-mile ascent was done in 3 hours, 45 minutes. A small girl gracefully played the piano once it had been safely deposited on the mountain top (*Ain't No Mountain High Enough,* possibly?).

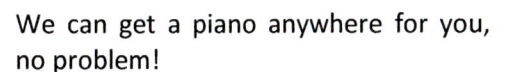

We can get a piano anywhere for you, no problem!

PIANO REMOVALS/REMOVERS: ADVICE AND POINTERS

Pianos vary in weight, size, value and sentimental value, so do the logistics when it comes to piano removals. At its simplest level, moving a small upright piano of low value, ground floor to ground floor, could be done by two level-headed and reasonably physically fit adults (remember to have protective blankets to cover the piano with, and ensure panels and lids, if not removed for transit, are locked or wedged closed so they don't fly open). In other situations, the enthusiastic removers have not always considered potential problems until they have met them en route, so plan ahead and be careful! The castors (little wheels) on most pianos, for example, aren't normally designed to be substantial enough for wheeling the piano down the street, they are likely to break or make the piano uncontrollable.

As some pianos can have considerable value, it makes sense to use a removal firm with the right experience and expertise. On the whole, you get what you pay for, so if it is an extremely valuable instrument, a removal firm experienced with national or international removals should be considered. They would most likely be a member of an organisation such as BAR – British Association of Removers. Moreover, in some cases the company would own quality vehicles with special suspension, in-vehicle security (cameras/sealable doors) and even climate controllable interiors (important for concert instruments where it might not have much time to acclimatize to its new setting. Additionally, pianos taken outside and placed in a very cold or damp vehicle are likely to suffer from rusting strings and sticking notes). If it is not a straightforward move, experienced personnel would also take the trouble to visit the client to inspect the piano and location before giving a quote. They would, of course, have the right amount of personnel needed and specialist equipment, such as blankets, crates, shoes (skid boards) and trollies. That said, such a service has to be paid for and is not needed by everyone. If your move is more local and for a less valuable instrument, consider your options.

If the removal firm has a website, check it out, look for reviews and/or use local knowledge by asking about reputations for different firms. How long has the firm been trading, are they well established? If they are really cheap, why might this be?

Whether a large or small firm, it might be worthwhile recording the removal on camera – just so long as you keep a distance and don't become a hindrance.

Is the firm a general removal firm or piano specialist? You might not need a piano specialist, some general removal firms are quite capable of moving pianos too (but perhaps not if it's a matter of removing window frames, using a crane or exporting a piano). With much smaller, local firms, they can often be far cheaper. Occasionally, however, they will take on a job simply because they need the work even though they have little experience of moving pianos. There could be a slight increase in risk, for example: they may have to manhandle/lift the piano up on to the back of a lorry where piano specialists would have a ramp and/or controllable tail-lift to get the piano on and off the lorry more safely and with relative ease.

It's worth checking to see if your job is the only one being carried out on that day, or will they be delivering other pianos too? This might involve potential delays; you should also know if the piano is being kept in the vehicle overnight. Similarly, it would be wise to confirm who will be actually moving your piano: sometimes a large firm might take the booking but use a subcontractor to carry out the work.

Well-established removal firms can arrange insurance of the piano during the removal process, smaller firms might not be able to do this (so check carefully to see who is responsible for insurance of the piano during the move).

Home insurance doesn't normally cover having a piano moved. When taking out insurance for piano removal, you might have to put in writing the approximate age and value of your piano, so make sure you have some idea. Find those receipts and scan them, or perhaps get your local piano tuner or dealer to give a valuation of the piano. Also be clear about what is being insured. If, for example, some expensive flooring is damaged during the removal process, is that (and walls etc) covered?

Removal firms often remark that people underestimate the size of the piano and how much room it takes up; measure height and length accurately. The length of a grand is measured with the lid closed, but from in front of the centre edge – allowing for the keyboard jutting out beyond the length of the top lid – and right down to the rear tip. Have such measurements and the model's name (also model number/series if it has one) handy when first dealing with the removal firm as they might already be familiar with the type.

Small children and pets should be kept well out of the way, ensure that obstructions such as electric leads and sofas are kept clear of doorways; allow room for the piano to be 'swung round' at an angle to negotiate certain doorways. Tables and other items might need to be removed from hallways <u>before</u> the arrival of the team.

Be clear about other impediments when discussing the removal: are there any stairs? How many flights? Are there stairs at one end of the removal or both? Is there a lift? If there is a lift, does it have a weight limit? Some firms will include a stair crew in their quote (so making the quote more expensive but minimizing possible damage to the piano, the surroundings or personnel).

It can be useful to send the firm photos of the piano and premises beforehand. This also helps to confirm with the firm if there is or isn't any existing damage to the piano's casework, it also might be an opportune moment to enquire if they have any satisfactory arrangements in place to rectify scratches, chips etc should they inadvertently occur during the removal process. Sometimes, a firm can be more fairly judged by how they react and respond when a removal hasn't gone quite right – are there excuses and blame or a concerted effort to help put things right?

Parking can lead to delays, so if possible try and secure adequate space for the removal lorry on the day of the removal (at both ends if necessary).

If the piano is being delivered for a concert, it helps if there is one 'named person' used as a point of contact and to liaise things smoothly (if it is a hired piano, does it come with a good stool; also, what tuning arrangements have been made?). In an ideal world, allow for possible delays and problems. For example, unless it is a very local delivery, the piano should be delivered a day or more before the concert, not on the day itself, this is not just to allow for slight risks such as human error, severe weather conditions, vehicle breakdowns, protests, strikes or similar, but other things need to be considered also. For example, will the piano be needed for a rehearsal and will that take place on the day of the concert or the day before? Have you allowed time for the pianist and piano tuner/technician to get access to the piano? If the piano can be delivered before the day of the concert, it also has a better chance of becoming 'comfortable': acclimatized to the new environment before being used. (Pianos moved into new homes should be left about a month to acclimatize before being tuned; a desirable temperature for pianos is around 21°C.)

Importing/exporting a piano? Modern pianos will not have ivory keys due to world legislation banning their use. However, there is legislation in place concerning older pianos with ivory keys, so ask your local tuner or dealer for advice. In some cases you might need a CITES permit to ship an older piano in or out of the UK (CITES: Convention on International Trade in Endangered Species; some removal firms can arrange this or else visit www.gov.uk/guidance/cites-imports-and-exports).

ROYALTY

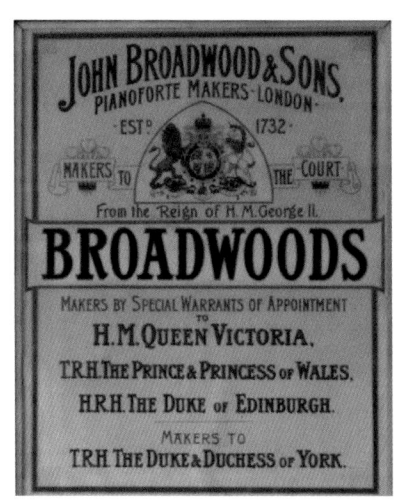

Members of the royal family were early piano enthusiasts even before Queen Victoria and Albert became 'devout' keyboard practitioners, though Victoria herself grew up just at the right time, when the evolution of the piano was transforming it into a more sophisticated and durable instrument and with a much wider range available.

There are approximately seventy pianos in the royal collection and dispersed among the royal residences. The earliest piano suppliers, however, before Broadwood and other well-known makers became more fully established, are largely forgotten names today. One of the earliest pianos supplied that is still in the royal collection is an upright by Rice & Co., who worked from 11 Golden Square, London (and had a side-line in merchandizing coal). The pristine Rice 'upright grand' was purchased by the Prince of Wales (later George IV) for £680 in 1808.

A grand by Isaac Mott followed in 1817 (he too had a side-line, in a distillery). As with most other royal piano suppliers, both Jones and Mott wasted no time in adding such words as 'Supplier to the Royal Household' to their literature. Having supplied one grand piano to George IV, Mott soon after managed to sell eleven more to English noblemen (he had, at one point, set up his piano business in Pall Mall with his cousin, one Julius Caesar Mott). After the Prince of Wales had become George IV, the entrepreneurial Mott was able to advertise as 'Pianoforte Makers to his Majesty' (he produced a guide to playing the pianoforte in 1820, titled: *I.H.R. Mott's Advice and Instructions for playing the Piano Forte with Expression and Brilliant Execution*).

Broadwood's, under the original ownership of harpsichord maker Burkat Shudi (friend of Handel), supplied a harpsichord to the Prince of Wales in 1740. After John Broadwood married Burkat's younger daughter, the business later became J Broadwood & Son. It supplied a grand to George IV in 1821, having become a regular royal supplier. Well over a century later, it gifted a grand piano to Prince Charles and Lady Diana as a wedding present.

Left: An 1821 grand made by the largely forgotten maker, Thomas Tomkison. The renowned maker had his workshop in Dean Street, Soho and supplied a square piano to Admiral Nelson, and pianos to royalty here and in Spain and Prussia. The grand was bought by George IV and kept in the Royal Pavilion, Brighton. Having gone out of royal ownership much later on, the Tomkison 6-octave grand fetched £62,000 at auction in 2017 and was returned to the Brighton Pavilion. Tomkison had known the painter JMW Turner since boyhood, as both their fathers had shops in Maiden Lane, Covent Garden.

Right: An 1839 Jean-Henri Pape square piano on display at Victoria's former holiday residence, Osborne House, Isle of Wight. Its walnut case is inlaid with ivory. German maker Carl Bechstein learnt his early piano craftsmanship with Pape in Paris.

Records for the 1930s show Buckingham Palace had seven pianos, while Sandringham had four.

Below: A very early receipt for a 'grand forte piano' priced at £102.7.6 supplied to George III by maker Robert Stodart in 1786. V&A archive evidence reveals that Robert, founder of the company, was essentially a piano tuner-builder who also taught music to princesses – the King's daughters (he is credited with coining and patenting the term grand piano). Below, right, shows Victoria, Albert and offspring. The composer Mendelssohn became a friend and frequent visitor, sometimes accompanying Victoria in song.

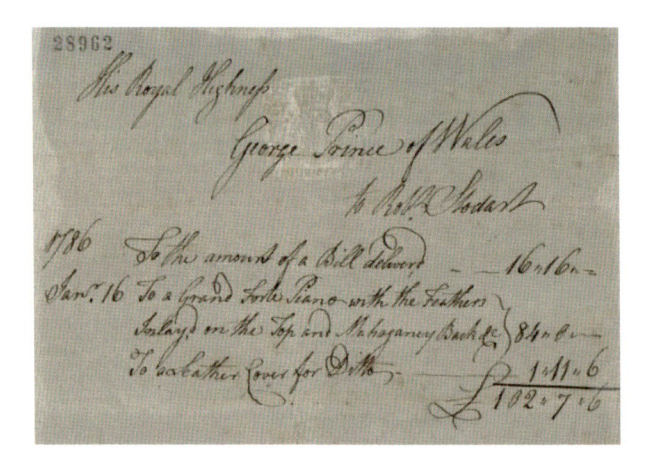 

The gold-leaf grand piano with incurving cabriolet legs shown below was purchased by Victoria from the leading French firm of the nineteenth century, Erard, who made pianos for the French nobility, including Louis XVI and Marie Antoinette. Erard had an excellent reputation, the composer Beethoven had earlier sought out and bought himself an Erard grand; the company was also renowned for its harps.

It was not the first Erard owned by the royal family, but this special model was purchased in April 1856. Such firms as Steinway and Bechstein were mere little-known upstarts at this stage, with the Erard firm having around half a century of

piano-making experience under its belt (they had factories in London and Paris). A closer look at the interior of the gold Erard (see overleaf) reveals it has iron frame bracings, so the design has moved on from the earlier wooden-framed fortepianos. On the other hand, it is an underdamper – grands after this time and to this day have dampers that rest on top of the strings rather than the dampers being pushed up on to the strings from underneath. On playing the Mendelssohn piano concerto number 1 in G minor on the instrument at the Proms in

2019, Sir Stephen Hough commented: *There is an intimacy, and human quality, to this piano's reedy timbre which really does feel like something from another era … Its shallow action and smaller hammers require the pianist to play in a totally different, less muscular manner.*

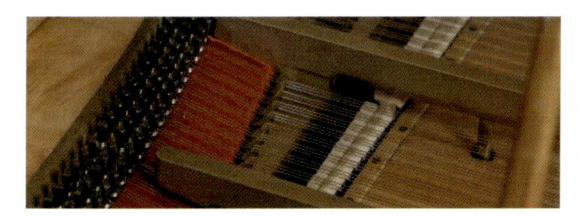

The restored piano was loaned by HM the Queen for the Proms concert, the first time the piano had ever left Buckingham Palace. As an encore, pianist Sir Stephen Hough played Chopin's Nocturne in E flat major.

Founder Sébastien Érard (1752 – 1831) was born in Strasbourg, where his father worked as an upholsterer. His father died when Sébastien was only sixteen, so he moved to Paris and worked for a harpsichord maker. It would seem his employer and fellow workers were jealous of his skills and ingenuity, as he was sacked. With the support of the Duchesse de Villeroi, he set up his piano making business in a hotel room and built his first piano in 1777 (later spending some time in London to escape the French Revolution; at this time and after his death, other family members were also involved in the business). There still exists today a rue Erard and Erard Salon in Paris.

Above: German piano manufacturer C. Bechstein has traded since 1853. In 2013 it commemorated its 160th anniversary by building a unique Louis XV cased replica of a gold grand piano that had been made and presented to the royal court of England. With rich gold leafing, delicate wood carving and elegant miniature paintings after the Watteau style, Carl Bechstein had been personally involved in the piano's design and construction.

Below left: The Queen Mother and King George VI. To the right are their two daughters, Princess Margaret and the late Queen Elizabeth II. Elizabeth began her piano lessons when aged eleven. Two of her children, King Charles and Princess Anne (Princess Royal), were taught by pianist Hilda Bor, who was a regular broadcaster for the BBC. A contemporary of concert pianist Myra Hess, who started the lunchtime concerts at the National Gallery during the second world war, Bor similarly organized lunchtime concerts at London's Royal Exchange.

 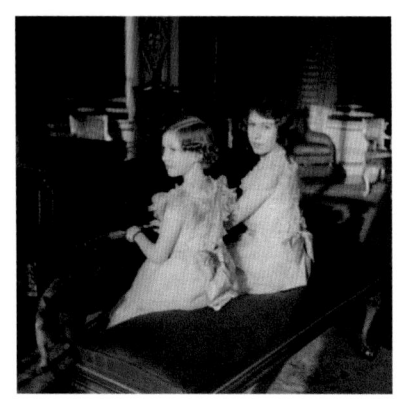

There are five royal warrant holder piano firms. The first royal warrant began with the Weavers' Company in 1155 (today, in addition to luxurious brands such as Bentley, are included Kellogg's and Heinz – their 'Catsup' later became Ketchup when marketed in Fortnum & Mason). The piano firms with royal warrants are: Gordon Bell Ltd, Aberdeen; John Broadwood & Sons Ltd, N. Yorkshire (since 1740); Period Piano Company, Kent; Steinway & Sons, London; Wilson & Son, Edinburgh.

S

STRINGS AND SOUNDBOARD; SITTING POSITION; STEINWAY; SILENT FILMS; STREETPIANOS; STAIRS, STEPS AND STREET CROSSINGS

Pre-1830s, most piano strings (or 'music wire') were made from iron and prone to break quite easily. Later and to this day, they are made of polished steel and can be tuned to a high tension. The average piano has around 230 strings; in the piano they are secured to a hitch pin on the iron frame at the end furthest away from the keyboard, with the other end being coiled to a tuning pin seated in the wooden wrest plank. The steel strings might all look the same, but if one breaks, a replacement of the correct gauge must be used, as the strings get thinner (and shorter) the higher up they go into the treble. The vibrating part of the top treble string is only a few centimetres long, and at the bass end, the first string is over one metre long. Each bass string is made thicker by having a copper winding wrapped around its inner steel core. This enables the string to emit the lower pitch needed; if the bass strings weren't made thicker, they would have to be very much

156

longer to get the right low pitch (but the longer strings needed would make the piano an impractical length). Piano strings rust surprisingly quickly in damp rooms, so moisture and even human fingers should be kept well away from the strings.

In many ways the soundboard in a piano is the beating heart and unsung hero of the instrument. A piano without a soundboard would be like playing an electric guitar with no power on – you'd hear next to nothing. The soundboard is made from fast and tight growing strips of softwood – usually spruce – glued together and then varnished. It is actually convex in shape, not flat, with its crown (facing upwards) being towards the centre of the piano. It is really a large wooden diaphragm that amplifies the sound of the vibrating strings which are stretched across the piano and pass over the hardwood (maple) bridges. These bridges transmit the sounds of the vibrating strings directly on to the large expanse of soundboard, using a down-bearing. The soundboard itself sits within the piano's rim and on a shelf. Often not seen, on the other side of the soundboard are screwed wooden ribs to reinforce it and help maintain its shape and condition. On an old piano (or occasionally an inferior one) the soundboard may have lost its shape, flexibility or is not receiving the correct down-bearing to allow it to produce the desired quality and quantity of sound needed. (See page 7 for soundboard picture.)

SITTING POSITION

Perhaps they are damned if they do and damned if they don't, piano teachers regularly give advice about using the 'correct' sitting/playing position. On the other hand, certain top pianists have always broken the 'rules'. Glenn Gould, for example, always sat very low, while Horowitz held his wrists low and used flat hands (in other piano genres, some even stand up or adopt more of a karate-style attack of the keyboard). Of course, humans are not robots and many are not, physically, a 'regular fit' (to a large extent, Horowitz could only play in his unique style by having the action altered – an action that would not have suited most other pianists).

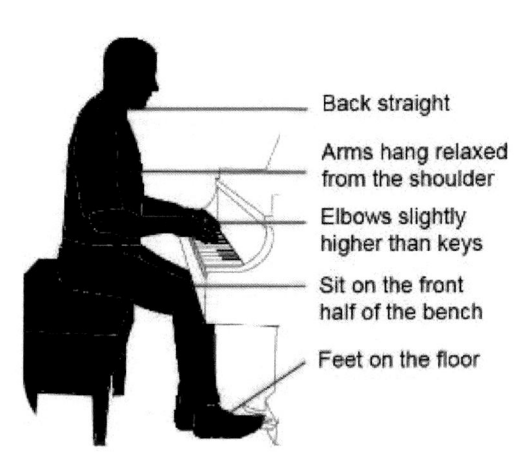

Back straight

Arms hang relaxed from the shoulder

Elbows slightly higher than keys

Sit on the front half of the bench

Feet on the floor

American concert pianist Charles Rosen (a former pupil of Moriz Rosenthal, who had been a pupil of Liszt) has quite a lot to say on teachers, the position of the hands and sitting position (the diagram is not his): *...almost all books on how to play the piano are absurd, and that the dogmatic system of teaching technique is pernicious. (Most pianists, in fact, have to work to some extent in late adolescence to undo the effects of their early instruction and find an idiosyncratic method that suits them personally.)*

He went on to say: *Not only the individual shape of the hand counts but even the whole corporal shape. That is why there is no optimum position for sitting at the piano, in spite of what many pedagogues think. Glenn Gould sat close to the floor while Arthur Rubinstein was almost standing up. It may seem paradoxical that some pianists spend more time choosing a chair for a concert than an instrument; the piano technician at the Festival Hall in London told me that the late Shura Cherkassky decided on the piano he wanted in five minutes, but spent twenty minutes trying out different stools. The height at which one sits does affect the style of performance...*

He wrote over two hundred piano pieces, also piano concerti and operas, but the pianist and composer Frédéric Kalkbrenner is a rather unknown name today. He was a friend and teacher of Chopin but was also involved in piano manufacture (Chopin dedicated his first piano concerto to Kalkbrenner).

Kalkbrenner lived in London from 1814 to 1823, where he both taught and performed. He made himself wealthy after developing an invention by Johann Bernhard Logier. Called the chiroplast (or hand guide), it was a wooden contraption where a rod of wood was placed in front of the keyboard and the forearm was

supposed to rest lightly on it, the idea being that it would encourage the right kind of muscle action and strength in the hands but suppress those in the arms and elsewhere. Because he had taken a patent out on it, the chiroplast proved to be a lucrative business decision (partly because both Logier and Kalkbrenner promoted the invention's benefits and used them in their piano schools; it was also a method taught to composer Camille Saint-Saëns when he was a piano pupil).

Right above: An unusual playing position! – see MOMA Crazy Piano Player on YouTube.

STEINWAY

There have been numerous detailed books and films on the history of the Steinway piano company and how they are made, so here I shall not focus too much on such things as statistics, medals, patents and model types.

The first Steinway pianos weren't made by Steinways! Nor did the firm of Steinway and Sons, founded in 1853, make grand and upright pianos as we know them, and the family name wasn't Steinway. Actually, the family had been making pianos for two or three decades before 1853, only in Germany, not America, so it's best we backtrack...

The family came from Seesen, a small rural town slightly north of central Germany (Lower Saxony, about 78km south of Hanover). The founder of the firm was one Heinrich Engelhard Steinweg (1797–1871, depicted on the previous page). Orphaned at fifteen and almost penniless, Heinrich became a cabinet maker after leaving school and was later apprenticed to an organ builder. After marrying, he bought a large family home which included a workshop, where he started making pianos in the 1820s (almost all square pianos, as would be the case when the family first established their business in America).

He sent one of his sons, Carl, over to America in 1849 to scout out the piano-making business there (it also allowed Carl to avoid joining the army); the following year fifty-three-year-old Heinrich immigrated to America with his wife and seven of his nine children. He and his sons worked for other piano builders initially before using a loft at the back of 53 Varick Street in Lower Manhattan. While his son Theodor continued to make pianos in Germany (with Friedrich Grotrian joining him as a partner in 1856, making Grotrian-Steinweg pianos), the Steinweg family made pianos from 1853 under the anglicized name of Steinway. This name was used on their pianos as it was thought it might be a more marketable name in the US (the family didn't officially adopt it until 1864). Around a year later they moved into larger premises in Walker Street and an outstanding reputation began to flourish.

 The first piano made by the family in America was numbered 483, it was given this number because they had already made 482 as a family firm in Germany. This first American model, a square – and the first to carry the Steinway name – was sold for 500 dollars to an American family. After a change of ownership and being displayed in the New York Museum of Art, it came to the museum in Seesen (and is shown left). As can be seen, Steinway refined their squares, also making them larger and, with metal frames, more durable.

In the earliest days the company's workforce consisted mainly of German émigrés, so the language spoken in the factory was almost exclusively German. It didn't take long, however, for the company to expand and build a new factory. This was in Park Avenue (from the 1860s), it had a capacity for 350 men who initially turned out around thirty square pianos and five grands a week. Also, in time, the company would be granted 139 patents, win many awards and was granted a royal warrant from Queen Victoria in 1890. Bechstein had beaten them to it, so to speak, but the company was ahead of the game in sales and marketing. Perhaps a hallmark of the company is not only the quality of their instruments but its creative ways for new marketing. For example, as early as 1866 it opened its first Steinway Hall. The Manhattan concert goers had to walk through the showrooms to gain access to the concert hall, which was said to have a seating capacity for about 2000 people. Twelve years later, the company opened a Steinway Hall in London's rather coveted Wigmore Street (moving to George Street in 1924). Interestingly, although Victorian Britain was in the midst of 'piano mania', with numerous very successful British piano makers building pianos for both the home and overseas markets, Wigmore Street had at least five key German makers with showrooms and/or concert halls by 1900: Steinway Hall, Bechstein Hall, Grotrian Hall, Blüthner House, and the Ibach piano company (founded in 1794, Ibach was older than all of them); but English maker John Brinsmead had moved into the street before most of the foreign rivals.

Following the illness and death of founder Heinrich Steinway, his eldest son Theodor came over to America to help run the company (he sold his shares in the Grotrian-Steinweg business). In just a few years, however, he found himself back in Germany running a new Steinway factory in Hamburg. This decision had been reached in a bid to gain easier access to the European market but also to avoid high European import taxes. Another son, William, took over the management of the US company; he became a leading and influential figure in business.

William was closely involved with establishing a company town – Steinway Village – in Astoria, New York for the workforce. Only remnants of the village exist today as the area has long since been redeveloped. In its heyday Steinway Village, built on scrubland, rundown farms and swamps, contained all that a small community of workers would need, including housing, gardens, a church, library and medical facilities. Although William Steinway had a certain philanthropic disposition, he was also motivated by wanting to get the workers away from what he considered bad influences, particularly union activity and those who wanted to strike and cause disharmony among the workforce. The Steinway company, at this time, as could be seen in other businesses, suffered a great deal from workers' strikes.

In the early 1870s Steinways moved from Manhattan to a new site in Astoria, New York (which had easy access to the East River). In time, the entire workforce lived in the company town (and were encouraged to buy their own homes). Perhaps it was quite fortuitous that at least one of the founder's sons (if not all) could speak English. When a pupil in Germany, young William had attended a go-ahead school where English was taught. While the original Steinway workforce had been nearly all German émigrés, most of the workforce today also originate from other countries and cultures. Although now occupying new factories, the company still operates from the same site today.

Like other companies, it would have its lean times, particularly during the 1930s, yet across the decades the company would go on to attract leading performers, from Paderewski, Anton Rubinstein and Arthur Rubinstein in the early days, to Van Cliburn later on and, most recently, Lang Lang – but also widening their market to include Steinway artistes such as Billy Joel (the company, incidentally, in more recent years even produced the Lang Lang Steinway, marketed towards its younger pianists). The New York factory produces around 250 uprights per year and 1000 grands (each grand takes nearly a year to make; black/ebonised grands have proved to be the most popular colour requested by customers), with the Hamburg factory producing a similar number. The Steinway piano has become synonymous with the concert hall and the world stage, a 'must have' brand for many.

When Steinway Village existed, some of the workers' names were used for some of the road names. Founder of the company, Heinrich, left his children a large estate valued at an estimated $500,000. By 1881 William Steinway had become a millionaire. As with other major piano makers, for a while Steinways had to contend with lesser-known companies using very similar names, Steinman & Sons being one example, or lesser-known manufacturers claiming to make pianos along the same lines as the Steinway company.

Above: Production at Rondenbarg, 1923. In wartime, the Hamburg factory went over to making bunk beds and rifle stocks, while the New York factory made gliders and coffins.

SOME FURTHER FACTS AND KEY DATES

Seesen, where founder Heinrich made his first pianos under his original name, Steinweg, was a rather insignificant place though did have a claim to fame. Not so well known today, the leading German violinist and composer Louis Spohr was about thirteen years older than Hienrich and spent his childhood in Seesen. Upon becoming a musician of high repute, Beethoven visited his home to practise his *Ghost* piano trio with Spohr (the violinist wrote that the piano was out of tune and that Beethoven's playing was harsh or careless). But since around the year 1820, violinists everywhere owe Spohr a debt of gratitude, for it was he who invented the chinrest they have used ever since.

Family lore has it that Heinrich won a bronze war medal! It was for 'bugling in the face of the enemy': he fought against Napoleon in the Battle of Waterloo and was the lone bugler who signalled the charge. Reputed to have a very good musical ear, in his barracks he was known to make instruments such as mandolins and zithers, and to also play along with the military band.

In 1854 Steinway built two pianos per week (there were numerous other German communities and piano firms established in America around the same time).

The first Steinway Hall opened on East 14th Street in 1866 and had space for around 2000 people (the Hall was also the home of the New York Philharmonic orchestra for 25 years). When author Charles Dickens arrived on tour, he delivered readings of his works there. A modern-day media man – musically – is Steinway fan John Williams, composer of music for *Star Wars* and other well-known films. The owner of three Steinways, of his latest, he said, "I quite fell in love with this one."

In 1867 it was estimated that 90% of American pianos sold were square pianos. In 1884 the first modern-type Steinway model D concert grand was completed. Soon after, in 1888, Heinrich's son, Theodore Steinway (originally spelt Theodor) entered into an agreement with G. Daimler of Mercedes to manufacture parts for the American Mercedes automobile and also marine engines. This arrangement lasted at least until 1906.

During the Second World War, Steinway Hall was draped with American flags to counter some competitors' erroneous claims the company was pro-Nazi.

Heinrich's great-grandson, Henry Ziegler Steinway, having graduated from Harvard University, joined the company in 1937. He would be president of Steinway & Sons from 1955 to 1977, but also the last family member to have any direct involvement with the company. Bought by CBS in 1972, more recently the company came into the ownership of investment firm Paulson & Co. Inc.

The great pianist Josef Hofmann had rather small hands and could not stretch more than an octave. Steinway actually built a piano for him with slightly narrower keys (for some reason he preferred Steinways which had only two pedals, not three).

1988 saw the 135th anniversary of the company and presentation of the 500,000th instrument during a gala at Carnegie Hall in New York. Eight years later, in 1996, Selmer Inc. changed the official name to Steinway Musical Instruments Inc. Shares were floated on the New York Stock Exchange. The Steinway share was quoted on Wall Street under the abbreviation 'LVB' (Ludwig van Beethoven).

Buried under America's La Guardia airport are the former beaches and amusement park built by William Steinway and partner. It was once a highly popular resort.

The materials and hammers used in Steinway pianos made at their factory in Hamburg, are not the same as those made in America. There are pianists and piano technicians who claim German and American Steinways are not the same, with those coming out of one factory being better than those coming from the other.

Steinway pianos started life in Germany, they were made by Heinrich Steinweg (as Steinweg pianos). The family later on set up their new business in New York, but after also going into partnership with Friedrich Grotrian in Germany, the pianos made there were named Grotrian-Steinweg, and they are still made in Germany today. After numerous lawsuits over the years, the Grotrian company (now Chinese owned) is forbidden to use the name Steinweg on their pianos in the USA.

SILENT FILMS

Watching silent films (or movies) could be an exciting prospect, particularly after the 1909 cinematographic act, which introduced safer and more pleasant venues. Occasionally, grander cinemas might have had an organ or even an orchestra, but for many the mainstay was the house pianist. Little did these musicians know that in only two or three decades their much sought-after talents would no longer be required, for technology came in the form of 'talkies'. From the late 1920s, films had sound – music and dialogue (and soon came in colour too) – so pretty promptly thousands of silent film pianists became silent themselves, laid off permanently.

For the pianist, the screen captions, which cinema goers were well used to skip-reading, were cues to which they had to react. Versatility was needed to pianistically create atmosphere and emotion to suggest a chugging steam train, an angry exchange, imminent danger, a spark of romance, escape or even a motor-car smash! It was all down to the cinema pianist to get what they could out of the piano – strident chords one minute, the next perhaps a snippet from a well-known classical piece – to enhance the experience of the punters. The Marx Brothers, Chico and Harpo, had once been cinema pianists, but sadly the cinema piano would soon fall silent for all (though pianos could still be heard loud and clear in the pub).

The 1909 act made way for venues being licensed, which included having sufficient exits. Also, there was a risk of fire because the actual film was made from a highly flammable material so there were rules about where the projector was kept. There was once plenty of work for cinema pianists. Even In the early days most towns had several cinemas and, before the arrival of the television (and Bingo halls later on), people often visited them more than once a week.

For the earliest keen cinema goers, attendance on a Sunday was not an option.

Saint-Saëns was possibly the first of only a handful of composers who wrote a score for silent film: the 1908 *L'Assasinat du duc de Guise*.

STREETPIANOS

A piano in the street? An odd sight perhaps, a target for both vandals and health and safety officials. Yet they have become an increasingly common sight since the first decade of our current century. Some people tend to talk of the demise of the piano ('no one plays them anymore') yet the humble piano seen out and about in public places has put the instrument in a new light. The public piano, free from discrimination of such things as class, age and race, has given countless people free access to an instrument, while others have described them as mood lifters and smile generators. The public piano has become a friend of the people, actually changing the behaviour of commuters and travellers, forcing – in a gentle way – many to pause, listen, enjoy, converse or be inspired to have a go themselves.

In some ways it is a blast from the past, for we forget that many a home and pub piano was turned out on to the street for communal singsongs, street parties and celebration of such things as royal birthdays, the New Year, jubilees and ends of war. (Actually, the 1911 England census has at least five people recorded whose occupation was 'street piano player'.) In this century, however, a newer concept and compound word surfaced: streetpiano. Streetpianos are pianos put out in public places for anyone to play, their locations include railway stations, airports and shopping malls. From keen inquisitive toddlers reaching up to the keyboard, to concert pianists passing through St Pancras station, with all the other players and non-players in the middle having a go too, the streetpianos – usually uprights but not always – get plonked and played by an enormous range of people.

The idea for streetpianos can be credited to British artist Luke Jerram (interestingly, a colour-blind artist). It seems an understatement to call him 'an artist', for his works are wide ranging, with indoor exhibitions, huge open-air ones and others with a contemporary and experimental edge. We mustn't forget 'topical', too, a recent sculpture being the 'Oxford-AstraZeneca vaccine glass', made to mark the ten millionth vaccination in the UK. It is under the loose label of 'live art' that Luke Jerram launched his 'Play Me, I'm Yours' scheme in 2007. Fifteen pianos were installed across Birmingham for anyone to play. The soundwave had started. Thirteen were put out in São Paulo, thirty in Sydney. 2009 saw thirty across London; by 2018 over 1900 streetpianos in seventy cities had been installed, and it is today a worldwide phenomenon (the Bristol-based artist told me he does also have a piano in the family home, which he enjoys playing, along with the guitar).

Left: Leadenhall Market at an unusually quiet time. This City of London covered market has a lot of history, having first been established in 1321. It was later partly destroyed in the Great Fire of London. In much more recent times, it was used as a location – Diagon Alley – for a Harry Potter film.

Rather like people who sit and watch their whirling washing at the launderette but without speaking to anyone else present, part of the concept and challenge was linked to breaking down the barriers, especially ones that prohibit people playing music in public places unless a special arrangement has been made. Surprise, surprise, it wasn't long before bureaucracy and officialdom in some quarters tried to

Image overleaf: Canada's David Pecaut Square, where forty-one special streetpianos were decorated by artists representing the countries participating in the Pan-Am Games.

stymie the very welcome, peaceful and harmonious idea of access for all. For example, although Birmingham council helped to finance the first project, they announced that pianos could not be placed or played on council owned public spaces. The City of London, similarly, although presenting fifty pianos across London in the Play Me I'm Yours spirit, to celebrate its golden anniversary, wanted organizers to obtain licences to use the pianos, the absurdity of which saw the matter discussed in the House of Lords. On the other hand, the scheme saw some of the pianos later being given to worthy homes after the three-week event, and people such as Sir Elton John donating a brand-new Yamaha piano for St Pancras International station.

One wonders if Jerram was aware of the 'Sheffield Piano' that was already in existence, it belonged to student Doug Pearman. He wanted to get it into his new flat in Sharrow Vale Road but couldn't get it up the narrow and steep stairs. Eventually giving in, he and a friend came up with the idea of erecting a tarpaulin over it, adding a sign which invited anyone to play it if they wished and seeing what would happen. It became popular, even had its own website despite Sheffield Council deeming it to be a public hazard. Bolted to a wall and slapped with many letters and notes of affection and support, the Sheffield Piano did go on to last a good five years but eventually became rather weather-beaten. Streetpiano projects large and small have continued in recent years. Yamaha launched Platform88 Underground Pianos, which saw numerous quality instruments deployed in London underground stations, where the special acoustics, along with the playing by stars such as Jools Holland and Jamie Cullum, could be enjoyed by people passing through.

Above: Jools Holland trying out a newly-placed streetpiano.

Left: A streetpiano in a green setting, on Jesus Green, Cambridge (yet even the MI motorway service station at Newport Pagnell recently added one!).

On one occasion, platform 17 at Clapham Junction, Britain's busiest railway station, had an appreciative audience after a keen but anonymous pianist started playing some songs. Each one was enthusiastically clapped as harried and stressed commuters, evacuated due to station overcrowding and a tube strike, decided to gather around and listen to an impromptu performance by someone taking advantage of the streetpiano near the platform. His musical good turn made the national news.

Not far away, South London's railway station in Herne Hill benefitted from a local resident donating an upright that became an almost overnight hit in the community. Known as The People's Piano (it is shown on page 168), the one thing one regular user wanted to be reunited with after a brief spell in prison, was the Herne Hill piano. Another locally well-known young performer who played the same piano is Anthony Bastion, he could often be seen and heard playing his own music but also classical numbers. He got into classical music when aged around seven. Not by being taken for piano lessons, but by hearing the classical music that was used on the computer games he was playing. Excerpts of Mozart and other composers spurred him on to take up the piano. (Although streetpianos were temporarily locked during the covid pandemic, demand for online piano lessons soared.)

Below: A streetpiano in Coburg, Canada.

There is a Herne Hill in Canada but the image shown overleaf is South London's Herne Hill and The People's Piano, a local focal point since around 2014. In covid times, even streetpianos suffered under lockdown, being padlocked, lacking a feeling of wellbeing, and being neglected...

During lockdown the sign on the People's Piano stated:

Apparently my keys can spread disease and we can't take any chances. So I'm sorry to say that I am CLOSED for the foreseeable future. I'll let you know when I'm up and running again.

Below right: As this sign on St Pancras International Station shows, some normality eventually returned (St Pancras Station employs someone to tune and maintain their two pianos every three months).

Left: Jamie Cullum wows the crowds on St Pancras Station.

Left: A streetpiano in Kiev, 2014, helps to bring some peace and harmony in front of riot police. Sadly, much bigger troubles were to come for Ukraine.

STAIRS, STEPS AND STREET CROSSINGS

Numerous places around the world have piano-themed public spaces. The Ruthanne Ludato memorial playground (left below) and piano shower playground are both in the USA.

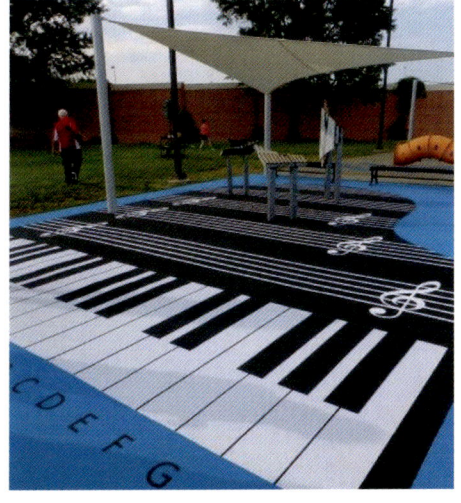

While the street crossings top centre and right, are in Manila and New York, public piano stairs can be found in Hangzhou, China and Stockholm, Sweden. For the latter, as the walker steps on one of the keys, it emits a sound. These Stockholm piano stairs were fitted for both amusement and to encourage people to take the healthier option. The external staircase, right, is in Valparaíso, Chile.

T

TONE; TRANSPORT; TUNING FORKS; TUNER (PIANO)

The tone of a piano is all about the quality of its sound, so can be a matter of taste or opinion: beauty is in the eye of the beholder. Descriptions for a piano's tone can also lead to unclarity, though it can sometimes be hard to put qualities of tone into words. Some descriptions I have heard are as follows, make of them what you will:

Uhm, cloudy, a rather tubby bass … Yes, a very silvery middle register … Overall, it's like honey, isn't it? … Such a muddy bass but the rest of it is so energetic! … No, it's all like cotton wool … It has colour … A bit of a wooden top end, don't you think? … Too shrill! … Rather treacly and needs more oomph … Superbly sonorous!

The problem is, to one person the sound is like cotton wool, to another it's warm and mellow. Whilst one pianist perceives the tone to be nice and bright, to another the same piano is harsh, shrill, glassy, brittle, even strident... One well-known concert pianist (source unverified) recalled about different pianos: *When I was about 12, I heard the great English pianist Myra Hess play Brahms's E flat Intermezzo (Op. 117, No. 1) on the radio. The radio in question was a small table model, but even through that single tiny speaker I could hear that Hess's instrument was different from any piano I knew. (It would be many years before I actually encountered a first-class Bechstein, which was what she was playing.) At 16, I touched my first Bösendorfer. I was enthralled by the sound, but I remember thinking, 'I will have to play differently on this to make it sound beautiful.' For me, it was always obvious that instruments played a role in determining interpretation; different pianos have different aesthetics, different concepts of beautiful or meaningful sound. It is a puzzlement why any pianist would want to play only one kind of piano: why would a wine connoisseur find the 'best' wine and drink only that one, or a car buff want the 'best' machine and drive no others?*

Closely linked to tone and performance are 'Regulating' and 'Voicing' (occasionally called 'toning' by some), which are covered on pages 145 and 186 respectively.

TRANSPORT

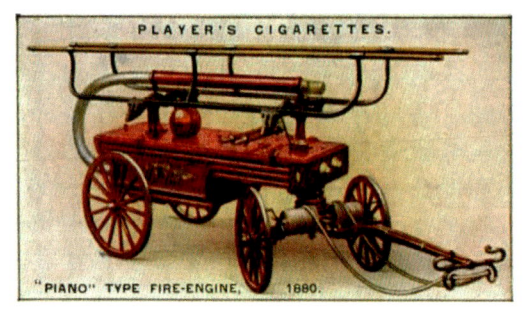

PLAYER'S CIGARETTES.

"PIANO" TYPE FIRE-ENGINE, 1880.

I'm not aware of any cars named 'piano', though there have been some with musical names. A few come close: Nissan Note, Honda Concerto, Hyundai Sonata, Kia Forte. There is the 7-foot Bösendorfer Porsche grand, on the other hand, impressively hand-crafted in Austria and designed by one FA Porsche (the lid opens automatically via a gas-pressured spring).

VARIOUS FORMS OF 'PIANO TRANSPORT'

Left: The interior of the Holland to New York cruise ship SS *Nieuw Amsterdam*, built in Belfast by Harland and Wolff in 1905. A few short years later they began work on the *Titanic*. As on that liner, the SS *Nieuw Amsterdam* had lounges for different classes. Although there were several pianos on board, it was only in the first-class lounge that a grand piano could be found (as seen here in the drawing room of the *SS Nieuw Amsterdam*. The piano is a Louis XVI styled Steinway).

A modern-day Holland to America cruise still runs and has a popular piano bar. The image, left, however, is from a 1930s cruise ship, a veritable floating palace, the SS *Normandie*. On its maiden voyage, from Le Havre to New York, the journey was accomplished in just over four days. For passengers travelling first-class, their cabins had the height of luxury, including a grand piano (third-class cabins had bunkbeds and little else).

Right: Not Murder on the Orient Express, but a piano instead (Agatha Christie did want to be a concert pianist but was ultimately too shy). Pianos have not only been transported by train for delivery but, as in this luxury Vienna journey (the 'Ritz on Rails'), have been regularly used for the passengers' entertainment.

Overleaf: A Return to water…

Enterprising musical couple, Rhiana Henderson and Masayuki Tayama launched their unique canal boat venture in 2020. Blending in well with the special community of London waterway users (among the more famous is Sir Richard Branson), they offer a range of musical options from concert cruises, residential holidays to afternoon teas. Indeed, Japanese-born Tayama is an experienced international concert pianist (to fit the new grand into their bespoke boat, the skylight had to be enlarged and a crane used to lower it in). Their narrowboat is named Rachmaninov.

To finish the water theme, the very first piano to arrive in Australia was a Beck square piano built in London. The property of ship's surgeon George Worgan, it was carried aboard the flagship of the First Fleet, HMS *Sirius*. The piano was used for recitals during the 252-day voyage and arrived in 1788 to a Sydney summer temperature of 104F. Worgan later gave it to Elizabeth Macarthur, to whom he had given piano lessons (the piano is still in existence).

TUNING FORKS

Tuning forks, in various forms, have been around for a surprisingly long time – from the Ancient Egyptian period and, left, Celtic times (the image is of an Abernethy Pictish stone from Perthshire). In musical terms, the tuning fork – often called pitch fork originally – as we know it today, was invented in 1711 in England by John Shore. An instrument maker and trumpeter of repute in the royal court of James II, Shore gave one of his tuning forks to court composer George Frideric Handel, it today forms part of the collection in London's Foundling Museum (established in 1739, Handel and artist William Hogarth were leading supporters of the Foundling Hospital, which is the UK's first children's charity and public art gallery). Handel wrote certain of his trumpet pieces for Shore (who also knew Purcell), but after damaging his lip, found he could not play to his former standard so turned his skills to playing the lute instead. It was perhaps partly to help him keep his lute in good tune, in fact, that he invented the tuning fork, which proved to be more reliable than the available wooden pitch pipes. Shore died in 1752 (no existing grave but one source gives his epitaph as 'died deranged').

There were once numerous firms in Britain making tuning forks, today only one firm remains, it is Ragg Tuning Forks, part of Uniplex (UK) Ltd, situated on Furnace Hill, Sheffield. They are makers of the Walker tuning fork and are still close to where the company started in 1841 – Nursery Works, Little London Road. Only a decade later a set of thirteen chromatic tuning forks (pitched to equal temperament) was exhibited at the Great Exhibition held at the Crystal Palace in Hyde Park.

While countless piano tuners of the past used a tuning fork to check the pitch of a piano (and some still do), tuning forks have had and continue to have other uses (the Ragg company sells around 35,000 a year). Most doctors have one as part of their medical equipment. They can be used to check hearing impairment or test nerve damage in patients with diabetes, for example. But they are also used by alternative practitioners, among them kinesiologists. A slight irony, perhaps, is the never far away issue over the widely used concert pitch of A440. It is used across much of the world but has had and continues to have its critics. Some believe the lower pitch frequency of A432 is a nicer and much more harmonic pitch, one with spiritual healing properties (and one initial Italian study published in PubMed in 2019 found that music tuned to A432 only slightly improved the blood pressure but did more significantly decrease the heart rate of its participants than music tuned to A440). Others also like the maths involved, as the frequency 432 and its neighbouring octaves contain easy round integers, for example: note A above A432 gives a frequency of 864, note A below 432 gives 216 and so on. So, outside the musical world, A440 was and still is considered by some to be a not so pleasant pitch (or less 'spiritual'), with some practitioners and believers feeling it is an aggravating pitch and disharmonious, somehow out of tune with Mother Earth.

Lastly, in the past, tuning forks were used to calibrate equipment in telephone exchanges. Even today, tuning forks are still used to calibrate speed cameras and radar guns. Theoretically at least, traffic officers are supposed to carry a tuning fork in their equipment to test that such things as speed guns are working accurately.

Formerly called vibrations or cycles per second, the scientific frequencies of vibrations became measurable in 'Hertz' (Hz) from 1930, named in honour of scientist Heinrich Hertz. He discovered radio waves and also that sparks produced a regular electrical vibration within the electrical wires they jumped between. There is a Hertz crater on the far side of the Moon named in his honour (Mercury, incidentally, has a Grieg crater named after the Norwegian pianist and composer).

The image left shows an unusual Ragg order made for Icelandic singer-songwriter Bork. The

company also produces a 9CT gold-plated tuning fork for gift or presentation purposes. Much earlier, at least one company made adjustable tuning forks, they had a metal slide attached to them so the pitch could be altered.

Right: An old set of three C forks, stamped with: Continental C 517; Philharmonic C 540; Medium C 530.

Left: Once in the ownership of Vaughan Williams, Beethoven's tuning fork is shown in its lined box (held by the British Library, as is his laundry list!), but poor Beethoven needed more than the primitive medical technology of the day to overcome his deafness (examples shown below). After watching Beethoven in a rehearsal for the *Archduke Trio* in 1814, the composer Louis Spohr said, "In *forte* passages the poor deaf man pounded on the keys until the strings jangled, and in *piano* he played so softly that whole groups of notes were omitted, so that the music was unintelligible unless one could look into the pianoforte part. I was deeply saddened at so hard a fate." Reputedly, it was his pupil, Czerny, who first noticed Beethoven's deafness before it became public knowledge. He wrote: *I also noticed with that visual quickness peculiar to children that he had cotton which seemed to have been steeped in a yellowish liquid, in his ears.*

Having repeatedly hammered away and left one of his pianos with broken strings (like a 'thorn bush in a gale' according to harp/piano maker, Johann Stumpff, his friend and faithful tuner), Beethoven asked piano maker Conrad Graf to build him a piano with four strings to each note.

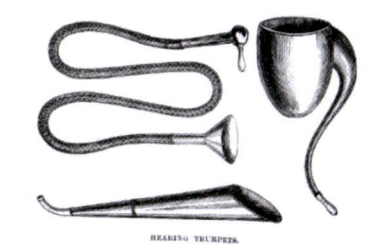

HEARING TRUMPETS.

TUNER (PIANO)

The piano tuner is the unseen artiste – both complimentary and true, but how did the first piano tuners get started, was there a City and Guilds-type course to enrol on? Obviously not, for the role of professional piano tuner evolved. There were no official tests, courses or recognized qualifications. Unsurprisingly, the first pianoforte tuners (as they became known) were the people who mostly built, maintained or played harpsichords and organs. As the first pianos to arrive on the scene were relatively small and lightweight, usually the 'square' piano type, although more demanding than harpsichords (with only one string per note), many of these existing instrument makers/repairers adapted their skills. Additionally, most harpsichordists were used to tuning their own instruments anyway as they could go out of tune even during the course of half a day (composer JS Bach was known to tune his own instruments). So along with owners and builders of harpsichords or organs, other instrumentalists, musicians and teachers, too, saw a gap in the market for the tuning and repairing of pianos. Thus, as the piano evolved during the Victorian era into an instrument still recognized today as the grand or upright piano, it wasn't uncommon to see adverts from organists and choirmasters which included the tuning of pianos to their list of musical offerings via local newspapers and church or musical publications. They were perhaps 'having a go' and teaching themselves. In some cases, it might have been more out of necessity as demand for the (evolving) professional tuner at times outstripped the huge sales of pianos across the country. Unsurprisingly, Teach Yourself books were published, and there were instances of bogus piano tuners too. That said, many must have realized that the most modern pianos, with around 230 strings tuned to a high tension, and with a rather sophisticated action, required skills that were beyond most hobbyists (not to mention being able to lay an accurate scale of equal temperament).

Of course, the rise in Victorian 'piano mania' meant there were now many firms building and selling pianos. A perusal of any trade directory at this time would show almost every single large town in the country had at least one firm either building or selling pianos. It was the larger piano builders, then, who were the first to apprentice their own tuners, simply because their own pianos needed tuning and, having been sold, it was often preferable to send their own men out to tune and maintain them. The bigger firms, such as Broadwood and Brinsmead set the standard early on. To begin with, and even after the Piano Tuners' Association was eventually established in 1913, it was the well-known piano firms where the independent tuners had first trained and worked which provided the necessary seal of approval. Offering a service and saying, for example, you'd earlier worked for Broadwood's or Cramer's, was the only sort of endorsement needed.

Regular work with a piano maker or retailer was enjoyed by many tuners who no doubt welcomed some security and perhaps a certain kudos if working for a more top-end firm. The 'small print' with these established firms, nonetheless, restricted tuners doing additional private work or, if later deciding to leave and become freelance, being legally prevented from carrying out their trade within a certain radius of the company (terms such as artisan, artiste, tradesman, craft/craftsman, profession – even the defining of 'class' itself – have always tended to be rather ambiguous and debatable terms as far as the term 'piano tuner' is concerned).

For the earliest independent tuners, when they were in demand but there wasn't sufficient local work to make it a worthwhile fulltime occupation, they devised their own 'tuning rounds' and would cover much larger parts of the country than most were prepared to do later on (despite there being better choices of transport). Communication was slower – most tuners would not have the use of a home telephone until the 1950s – so tuners would advertise in newspapers when they were about to visit the next town on their list. A prominent shop in the town would be used as the go-between (perhaps a book seller), with interested customers being asked to send a card with their details to that address. Upon arriving in the town, the tuner would then call at the proprietor and from there make the necessary arrangements to fit his clients in. Presumably he posted a card to the customer announcing his visit (trying to tie-in with the suggested times on the card, or perhaps he was free to turn up when available – the better off would nearly always have staff to let visitors and tradesmen in). Either the proprietor was obliging in accommodating the tuner's appointments system or perhaps they enjoyed a small commission for their trouble. Having 'done' that town, the tuner would not necessarily return to his home, instead proceeding to the next town either using his own horse/pony and trap or the stagecoach. From the 1830s there would be the faster and more convenient choice of the railway – would the tuner travel First or Second class (upholstered seats versus wooden ones)?

In recent times, tuners wanting work in piano showrooms or as 'outside tuners' for these firms have usually had to do a tuning 'test' where a resident tuner would check over their work and decide if it was satisfactory or not. For colleges and other institutions wanting the services of a tuner, it was always by recommendation and the name of the company or companies the tuner had previously worked for.

In the last few decades, tuners no longer keep their customer details on cards kept in boxes. Data bases are used, advertising and appointments are normally via websites and email. Overheads for materials, tools, expensive electronic aids, insurance, websites, accountancy fees, CRB checks, congestion charges, parking and other travel costs are in stark contrast to the running costs of tuners not so long ago. There is also the travelling 'electronic piano tuner', too, as the profession has been split into two camps (with some having a foot in both): the aural tuner and those who tune by using electronic visual devices – traditional v modern.

In time, polytechnics and colleges offered courses in piano tuning/maintenance. These altered in number, also in name. The London College of Furniture is one, it opened in 1964, and finally closed in 1992. Possibly the lead had been set by Claude Montal in France from as early as the 1830s. The son of a saddle maker, he became blind by the age of five but much later taught himself piano tuning, going on to write books on the subject and establishing his own factory. It was Montal who helped to set up comprehensive training schemes for the blind, particularly as they had great potential for independence and financial security. Thus, in England in 1872 the Royal Normal College was established in South London's Upper Norwood. Its resident blind boys were offered courses in music, the organ, and piano tuning. Its tuning courses would last until 2008 (by that time it had long been in Hereford – where former Home Secretary Lord Blunkett was a pupil, though he never took any tuning courses there). Currently, there is only one college in the UK offering recognized fulltime degree level courses in piano tuning and maintenance: in Newark. Not surprisingly, the occasional individual or newer small establishment has appeared, dubiously claiming their short intensive courses – some of only six months or less – will give individuals all the training they need to learn the craft and set up their own tuning businesses after completion of the (expensive) course.

Below: An early advert from an instrument/music seller, the *Caledonian Mercury* of 1796.

N.S. AND CO. beg leave further to intimate, THAT they lately engaged a TUNER of INSTRUMENTS from the House of Messrs. Broadwood and Son. Those who are pleased to favour them with their employ in that line may now depend on the most punctual attendance, and having their instrument done in the very best manner.

Left: Horse or horse and cart transport was once quite common for many piano tuners. Some at this time could turn their hand to repairing harmoniums and organs. In much earlier times, when the profession was still evolving, tuners not attached to a piano factory or music store sometimes had a second occupation or business interest as there wasn't sufficient demand for their services.

Certification added officialdom. The 1898 Brinsmead company certificate has no obvious sign of a grading system for new piano tuners, other than being rather briefly described as 'thoroughly practical'. Does number 117 mean the firm had trained that number of tuners by that year? Music dealers, when advertising for tuners, sometimes touched upon issues of dress, reliability and sobriety.

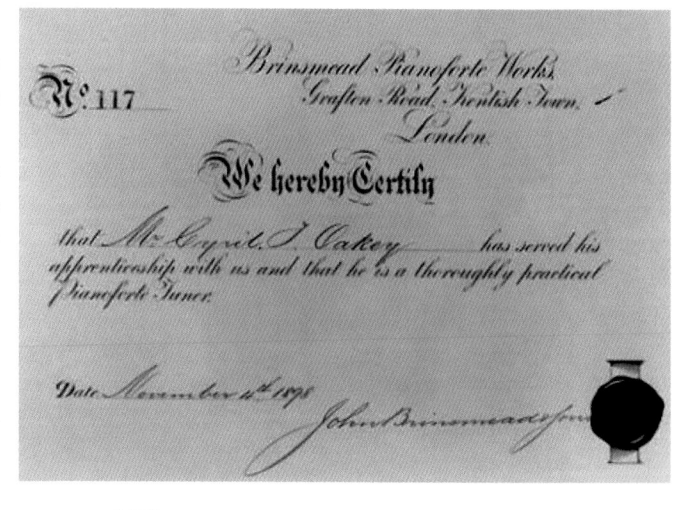

The Royal Normal College (later 'National' replaced 'Normal') was founded in 1872. The student in the left foreground appears to be fitting a new string. Mr. John Young, of Broadwood's, was in charge of the tuning department for thirty-five years at the college. After several years of instruction, the pupils gained experience by working in piano factories in London. Before entering the workforce, each student was required to pass an examination and obtain a certificate of proficiency from an expert authority on piano tuning. Mr. A. J. Hipkins, who had started with Broadwood's as an apprentice piano tuner at the age of fourteen and risen to become both their head tuner and a respected musicologist, was college examiner for many years. The tuning department was supplied with models of different piano actions and parts, which were renewed as new methods of construction were introduced by piano makers.

Right: There still exist numerous piano businesses with a family history, where the current owner was taught piano tuning by his father, who in turn was taught by his father. Founded in Grantham by William White in 1867, White & Sentance still trade today as a family firm in Sleaford. George Sentance joined the firm in 1897 (he had been apprenticed to the long-established London firm, Kirkman). Some piano retailers struggle somewhat because there are those who (rather unwisely) buy pianos from eBay and auction houses. The previously heard reason that a family can't have a piano in case it becomes a noise nuisance, no longer applies. Some come with built-in technology that incorporates a silent-play and record facility.

Piano Tuning Fees Across the Years

Taken from archives, newspapers and retired piano tuners' records, although there were minor fluctuations in fees across the country, it will be seen that for well over a century the average tuning fee remained at a stable three to four shillings. Occasionally, tuners put their fees up but some had reduced them at a later date (most square pianos had fewer strings so could be cheaper to tune). The pre-decimal shilling (1/- or 1s – 5p in today's money) was equal to 12 pennies (12d); 20 shillings (240 pennies) made one pound. A guinea was worth 21 shillings (£1.05 in decimal money). Be it a hundred years ago or today, there were and still are some tuners and companies who are surprisingly coy about their charges!

1772: London-based harpsichord builder, John Broadwood, tuned a harpsichord in Epsom for 5/-; after some experimenting, he began piano production in the 1780s.

1815: Music warehouse – grands 4/-, other pianos 2/6 (organ tunings from 10/6).

1829: W. Tierney, London tuner, grands 3/-, squares 2/6.

1851: London firm, grands 4/-, other pianos 3/- (skilled engineer's daily wage: 7/6).

1872: Erard company, 5/-.   New small Broadwood uprights sold for c. £20.00.

1875: Broadwood company, London area – cottage (small) pianos 3/-, grands 3/6.

1887: Cumberland area, pianos 3/- (fees always higher when more travel involved).

1900: Duffield, Belper, tunings 3/6 ('three and six' is equivalent to today's 42p).

1910: A school piano in Bedford, 5/- (tunings done through a music store, as this one was, tended to cost more – no doubt linked to commission and overheads).

1913: Halifax, 3/-; yearly contracts 10/6 (usually three or four tunings a year).

1917: Sheffield, 3/6. (1920 saw strike by piano workers, asking for 2/3 per hour.)

1928: Tuner employed by Pamment & Smith, King's Lynn, 3 tunings a year: 12/6. A London factory apprentice tuner was taken on and paid 7/6 per week.

1933: Harrods Ltd 4/6. 1938: E. Price Ltd, Yeovil, 4 tunings a year, £1.2.0.

1943: A Bedford farmer paid 6/- for a tuning (account shows 2 horses shoed: 17/-).

1947: *The Scotsman* advert: shop piano tuner, wages £5.00 to £5.10s per week.

1950: London uprights 7/8.

1953: Duck, Son & Pinker Ltd, Bristol, tunings 10/-.

1963: Prices varied, 9/6 to 12/- quoted in numerous sources.

1966: Fees were around 17/6.

1970: £1.10 to £1.17.6 (decimalization arrived the following year, 1971).

1972: Tuner working for Berry Piano firm, London, £2.75 upright, £3.50 for grands.

1975: London, average of £4.50. Harrogate apprentice chef wages: £15.00 a week.

1979: Grimsby area, £8.00 for a standard tuning, both uprights and grands.

1981: *The Times* consumer guide stated tuning prices were £9.00 to nearly £14.00.

1988: London, £17.50.

2002: Average of £35.00 across Britain.

2011: £40.00 to £45.00; 2014: Average of £50.00.

Tuning Fees 2023: Bournemouth £65.00; Norfolk £75.00; Blüthner Pianos £96.00; Samuel Pianos (London) £105.00; Markson Pianos (London) £90.00; Devon £60.00; Pianos Cymru (Wales) £65.00 to £75.00; Glasgow £65.00.

The image, left, was not the first piano tuner to feature in art (there is Frederick D Hardy's *The Piano Tuner* of 1881 for example), but *The Piano Tuner* (left) was painted in 1947 by the well-known American artist Norman Rockwell. The boy who posed for him was paid $5 (it was later gifted to the boy's parents). The tuner in the painting, it would seem, is blind, as he appears to have a collapsible mobility cane. It sold at a 2018 art auction for $2.7m. It had first been reproduced on the front of the *Saturday Evening Post*.

U

UNISONS; UBIQUITY OF THE PIANO; UPRIGHTS – see page 91

For around three quarters of a modern piano, each note has three strings (reducing to two strings in the bass and, for nearly the final bass octave, thicker single strings only for each note). For the notes which have three or two strings, the strings have to sound the same – as one – or else the note will sound out of tune (giving a vibrato effect to begin with). The unisons and octaves going out of tune can be quite noticeable to many people, leading to them wanting the tuner to visit.

UBIQUITY OF THE PIANO

Pianos get everywhere, even hung from ceilings, thrown into raging rivers or taken underwater via submarine! Britain's first touring caravan (left), pulled by two horses, set off in 1885 complete with a piano; Captain Scott later took one nearly to the South Pole. They have also been taken into the jungle, up mountains, and placed both in the desert and on ice floes. The images on these pages show how widespread the piano or piano imagery was and still is. A piano hasn't been taken to the Moon yet, though a story about a piano on the Moon was published in 2019. But there again the piano or piano tuner has also featured as the title, as a character or physically in art, literature (a blind piano tuner makes appearances in James Joyce's *Ulysses* for example), in song, on stage and in television plays, opera, film and ballet – *The Piano* (inspired by the award-winning film) was premièred by Royal New Zealand Ballet. One mustn't forget piano tuners depicted in saucy postcards and there's a 'Piano Tuner' perfume too.

Left: A grand turned into an upright? It had to be specially prepared so that pianist Alan Roche really could hit some high notes. The short concert took place in Geneva with a crowd of one thousand looking on. It was all part of the festivities to mark the re-opening of a museum.

The piano was and still is omnipresent, turning up in all sorts of places (a new initiative in 2022, for example, had one London borough providing a piano, free to use, in each of its libraries; Tunbridge Wells library currently has available for its members free use of a modern grand piano; there is also Channel Four's popular *The Piano* competition). The rectangular cake above was made by a Great British Bake Off contestant. There are also two plays shown here, *Piano/Forte* (Royal Court Theatre 2006) and *First Piano on the Moon*; complete with songs, the latter was put on at the Edinburgh Fringe Festival. Back in 1915, there was a piano related show song by Irving Berlin. Having had great success with *Alexander's Ragtime Band*, a line from Berlin's new song *I Love the Piano* included 'I know a fine way to treat a Steinway' (it was originally performed on Broadway with six pianos).

Above and right: A US submarine. The piano made it more homely?

Below: A glacier performance for Greenpeace. At the other extreme, regarding climate and locations, Egyptologist Arthur C Mace, who assisted Howard Carter with the Tutankhamun excavations in 1907, although living in rather basic huts, enjoyed the benefits of a Bechstein grand piano. Delivered by camel, it had been sent out by the parents of his wife, who had travelled out to Egypt with him. It was hoped the Bechstein would add to their creature comforts.

The upturned piano on page 184 was part of a display used in America to raise awareness (particularly among women) for heart disease. It was said that the odds of getting heart disease were 1 in 3, and the odds of being hit by a falling piano, 1 in 250 million.

Left: Polish composer Chopin. The bronze statue, which includes his muse and – symbol of Poland – an eagle, is just off Manchester's Deansgate, close to where he visited in 1848. Sculpted by Polish artist Robert Sobociński, it was unveiled in September 2011.

V

VIRUS; VC PIANO; VOICING

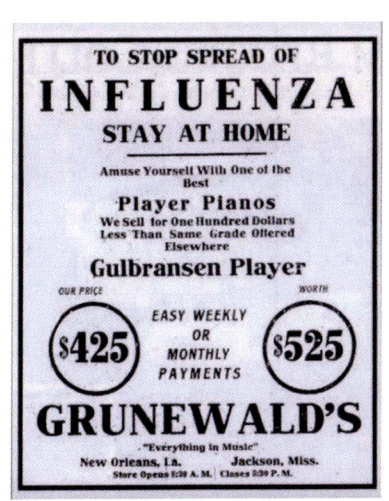

Linking viruses and corona, a Mexican pianist with a topical name is Claudia Corona, who lives in Germany though has played concerts throughout Europe. There is also a piece of music titled *Corona Suite* by Simon Peberdy (for euphonium, baritone and piano). The English piano makers Barratt and Robinson once produced a popular *Corona* upright. The poster shown left is of 1918, at the time of the serious Spanish flu epidemic (it cut short the life of composer of *Jerusalem*, Sir Hubert Parry). With one of the poster's subheadings saying *Amuse Yourself With One of the Best Player Pianos*, it uses the marketing angle of keeping safe but amused, and with a player piano you don't have to be a pianist!

VC PIANO

Bösendorfer are renowned for their VC range of concert grands (though they do also make upright pianos too). As can be seen, the company is very creative: the images here and overleaf show the special edition 280 VC concert grand. It was designed for Canadian Rapper Drake by architect Ferris Rafauli and designer Takashi Murakami.

Top of the Bösendorfer concert range is the awesome 9.6-feet 290 Imperial. Weighing in at 552kg, it has nine extra bass notes, giving a full eight octaves range (C to C). Extra notes are nothing new for the company, they first started making the wider compass Imperials in 1909. This came about after the composer Ferruccio Busoni needed more notes after transcribing JS Bach's organ works for the piano. He wanted to replicate the organ's lower bass pipes, but pianos at the time didn't have the sufficient keyboard range for it.

VOICING

In his writings, pianist Alfred Brendel has commented:

Concert grands of recent decades have progressively tended towards the harsh and percussive – or so it seems to me. (The great old pianists would have turned away in despair.) Pianos of the past displayed an inner resonance that gave the sound length and warmth. Yet even now it is possible to find, once in a while, a wonderful, magnificent instrument. Frequently, it has been monitored by one of the leading concert technicians. My collaborations with the finest exponents of this trade count among the happiest experiences of my musical life.

Voicing (occasionally referred to as toning in certain quarters) is something that can be done to the felt on the piano's hammer heads to alter the tone. It is not something that needs to be done on the average piano in the home whenever it is due for a tuning. Voicing is more often saved for good quality grands being used for recordings or concerts; that said, long term, the condition and shape of the hammers on any piano will alter after a lot of playing. This, along with the hammer felt becoming worn or too compressed over time, will change the tone of the piano or make the tone less consistent throughout. So long as the hammer felt is not too worn – not requiring a new set of hammer heads to be fitted – very occasional voicing is recommended as it will improve the overall tone on most pianos.

Different methods can be used for voicing. For example, the shape of the part of the hammer felt that strikes the strings can be altered/reshaped using sandpaper, or the felt can be pricked with needles to soften it if the tone is too harsh. Alternatively, the tone can be improved by hardening the hammer felt through ironing it or applying sparing amounts of acetone to the felt. In general, voicing of the piano helps to ensure the piano's tone is even throughout and that there isn't a note, notes or area of the piano which stand out for the wrong reasons (rather like having a choir where an individual's voice doesn't quiet blend with the others; alternatively, the bass section is too loud or the trebles are not pure enough...).

The subject of voicing can lead to many discussions and opinions. Earlier on I was careful to use the word 'improve' sparingly, as voicing can be subjective and words to describe the tone can have different interpretations, depending on who is using them. A piano may sound 'bright', but to another pianist the use of 'bright tone' means it is glassy, loud even or too shrill. Another pianist will describe the piano as having a 'warm' tone, whereas to someone else it sounds woolly and lifeless – far too quiet perhaps. So it's not always a matter of correcting the tone but adjusting it for individual tastes and preferences. Concert tuners will tell you that sometimes a piano is ideal for one venue but it can have completely the wrong sound when used at a different venue unless the tuner-technician sets to work voicing it (it's also worth bearing in mind that a piano heard in the showroom can sound quite different once it is in the buyer's home). Admittedly, perception of the piano's tone can also depend on the piece being played (and how), the mood of the pianist, the surrounding acoustics (with and without an audience), the warmth and mood of the audience, the weather, even the age and condition of the ears doing the listening...

The problem is, trying out a concert piano earlier in the day, without the audience being present or in some cases without the piano being surrounded by orchestral players, with different heating and lighting, and the concert performer being 'out of uniform' (possibly the piano isn't even in the correct position) can influence the pianist's perception of how the piano truly sounds. Concert pianist Charles Rosen noted he was quite taken aback at how different a note can sound even if the pianist is merely standing up or standing to one side of the piano (having the music-rest in position and raised up apparently also makes a difference).

Partly due to the many possible variables above, and sometimes preconcert nerves, some concert pianists can be unreliable and, dare I say it, are not necessarily always the best judge of how a piano should sound. Most piano tuner/technicians also know the adage well: tune the customer as well as the piano. It is not unusual for a pianist to perceive a minor fault with the piano and wanting it rectified or improved. In their absence, the tuner has worked hard and made the necessary adjustments. Only sometimes they haven't! They might have gone through the motions, but knowing there wasn't really anything amiss, they didn't want to undo their work but also not offend or upset the performer. Having returned later on and perhaps in a different mood or after a cup of tea, the artiste tries out the piano and expresses pleasure that the tuner-technician has corrected the 'imperfection': the piano is now in perfect shape and ready. Both artiste and piano have been tuned.

Perhaps the expression 'less is more' would be a useful one for tuner-technicians, as it is important they proceed with caution when voicing a piano. A concert piano doesn't belong to them, nor the artiste after that one concert, so altering the tone (or regulating the action for a different touch) too much might leave the concert venue or piano's owner with a piano that will no longer meet the general needs and tastes of most other pianists. Yes, genuine technical faults must be addressed, but I'm almost of the opinion that as most concert venues offer a high-quality instrument (top venues such as London's Wigmore Hall always have several concert grands for the pianist – often renowned visiting internationals – to choose from), the pianist should do the best with what they've been given. Most audience members are using their eyes to look at the performer and ears to listen to the actual music, they think much less in terms of whether notes on the actual piano have a bright, indifferent or mellow tone.

One example of the subjectiveness of a piano's performance and tone is that of a piano student being the rather brave subject of a filmed masterclass with Sir András Schiff. London's Royal College of Music had two identical Steinway grands on the platform for the masterclass, yet after a while the student asked if he could switch to the other Steinway because he preferred it. Pianist Sir András also subtly suggested that he felt the pianos, in his opinion, weren't really suitable for the Schubert pieces being played.

 Left: Significant wear and indentations from the strings (here, trichords – three strings to each note) can be seen on the face of each hammer. The hammer felt can be reshaped using sandpaper, but not too often. If the piano gets a lot of use, eventually a new set of hammer heads might be needed as, after reshaping, the originals would be too small and lightweight, having lost too much felt.

Israeli pianist Boris Giltburg wrote: *My dream piano has a singing, translucent sound, rich, varied, with a long decay; every note is rounded and bell-like. It has a broad dynamic range. Bass, middle and top registers are uniform in colour; there are no weak or unclear areas; nor are there any overly bright or open ones. Mechanically, the keyboard is 'even' (keys equally weighted); a touch neither too*

heavy nor too light, allowing full control over the sound. All of this unites into a whole larger than the sum of its parts that invites you to explore new areas and layers in the works you're performing.

You can infer the 'don't likes': a metallic or unclear sound that's flat and unvaried; a narrow dynamic range; an uneven keyboard, lack of character, and so on. Perhaps my demands seem exaggerated, but it is when working with the finest materials and tools that an artiste, in whatever field – whether music, painting or cookery – will achieve their best results.

WEIR; WEDDING; WREST PLANK; WARTIME

In architectural terms, 'piano' is used in the term *piano nobile* (Italian for 'noble floor' or 'noble level', also sometimes referred to by the corresponding French term, *bel étage*). It is the principal floor of a palazzo. This floor contains the main reception and bedrooms of the house. Moving on to hydraulic engineers, 'piano' comes into their vocabulary via a special weir known as a Piano Keys Weir (shown left), which is a new shape free-flow spillway.

WEDDING

Romantic Samuel, the groom, really did push the heavy Gaveau upright through the streets of Paris to meet up with his girlfriend at the Eiffel Tower, where he wanted to sing his wedding vows. (The Tower, early on, had a grand piano placed in the top floor apartment for its designer Gustave Eiffel.)

WREST PLANK

The wrest plank (or pinblock) is a plank of hardwood – typically beech – that holds each of the tuning pins to which are attached the strings. The steel tuning pins (often called wrest pins) are not threaded, each one is hammered into a hole in the wrest plank as tightly as possible so that it can hold the pull/strain exerted by each string tuned to a high tension. On early pianos the pins were oblong shaped; square-ended pins came soon after and allowed tuning hammers to be placed on the pin at a greater range of angles, making it much easier to manipulate the lever and pin. The wrest plank is a critical part of the piano's construction; after many years of tuning (and wear and tear), the holes will get marginally larger and the tuning pins will have less grip and become loose, leading to the strings slipping out of tune. One remedy is to re-pin the piano using a larger set of tuning pins.

Older and/or lower budget pianos may have a wrest plank made from one solid piece of wood; should cracks develop in the plank, they might get larger and larger, with the consequence of the whole tuning becoming unstable. Most modern pianos have a wrest plank made up of layers of wood bonded together, with each layer having the grain running at a different angle to its neighbour. Should a crack develop in one layer, it is unlikely to affect the other layers or cause problems.

Left: Can you spot the cracks on this upright's wrest plank that might make the tuning pins loose? Particularly on grand pianos, the wrest plank is not always easily visible, as it can be hidden under the iron frame. The frame and plank have matching holes to accommodate around 230 tuning pins.

Right: Notice how each tuning pin slopes back at an angle of around eight degrees (the iron frame is not shown). Piano owners should always try to avoid excessive heat and dryness. While dampness could lead to rusty strings and sticking keys, heat can lead to the wooden plank changing its natural shape, shrinking or expanding, with the possible consequence of the pins becoming less tight.

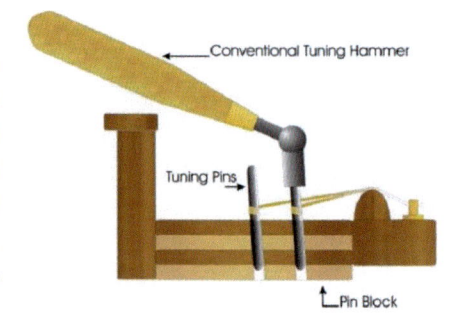

WARTIME

Pianos have proved to be remarkable morale boosters in times of war, whether during the Blitz or for serving soldiers. Even if in poor condition, many got extensive use (and with the rarity of available piano tuners, some of those in uniform who were tasked in putting them right, enjoyed the brief time off from the often tedious routines their fellow soldiers had to endure). Prisoner of war camps, in this country and abroad, nearly always had a piano and someone who could play it. Yet music and the piano were used in all kinds of wars. During the final years of the Cold War, Russian pianist Sviatoslav Richter embarked on a six-month road trip to the farthest reaches of Siberia, giving a hundred concerts in remote villages and towns.

As the images show, pianos helped to fix the morale of both POWs and serving soldiers (and also everyday people in the London Blitz).

Pianos had other uses, too. According to one newspaper report of 1916, an English accountant managed to escape from Antwerp and into Holland after the Germans took the city by being smuggled across the border in the case of a piano.

Although most piano firms stopped making pianos during the Second World War (instead often manufacturing other parts, particularly for aeroplanes), three were permitted to continue under a special government act. Two rival firms were moved to Welmar Pianos for the duration. It must have been rather odd for the older craftsmen from firms George Rogers and Sir Herbert Marshall & Sons sharing workshops with their trade competitors. On the other hand, elsewhere Knight pianos were contracted to manufacture pianos for the ENSA concert parties (author David Wainwright informs us that nine pianos were even sent ashore immediately after the Normandy invasion of 1944).

Left: Russian soldiers, Chechnya 1994.

Perhaps long forgotten now, though a few might still be in use in the schools and churches they were later donated to, the 'Victory Vertical' piano (or G.I. Steinway) was designed and built by that well-known company during the Second World War. As with other piano manufacturers, the common resources such as iron, steel and copper used in piano manufacture were badly needed during the war, so Steinway and other firms used their available materials and workforce to help the war effort. Steinway helped by making parts for gliders. But as has already been stated, the morale and welfare of serving soldiers had to be considered, and so it was that the special lightweight Steinway uprights were made and shipped out to places near and far where serving soldiers were stationed. The pianos were robust, each one having a secure wooden crate – partly because some were (possibly!) expected to be dropped out of planes. Water resistant glue was used, along with anti-insect treatment. They were also designed to contain one tenth as much metal as pre-war ones. Complete with carrying handles, they were reasonably portable and came in olive green, blue or grey – matching the service-type they were sent to.

Left: William Kuehl of the 10th Special Service Company tuning a piano in Guadalcanal in the Solomon Islands. Each piano came with sheet music, a book of instructions, spare parts and a tuning kit.

By the end of the war, around 5000 Victory Pianos had been made.

Below: Small, lightweight and ready to go.

In the grounds of his busy piano restoration business in Hiroshima, piano restorer and tuner Mitsunori Yagawa opened a piano museum in July 2021. Over the years, Yagawa, a second-generation atomic bomb survivor, has rescued pianos damaged from the 1945 atomic bombing. His mission has become to spread the word of peace to Japan's younger generations.

Initially, Yagawa had not been interested in any sort of peace activism and recalled that his late father rarely spoke of his wartime experiences. All that changed for his son after a piano was brought in via a group of atomic bomb survivors. It had been found in a house about three kilometres south of the A-bomb hypocentre. Japanese newspaper, *The Mainichi*, explains what happened:

The instrument's scars suggested it had been slammed against a wall by the blast. Yagawa was shaken by the stories of the damage done to the piano and how it had been preserved, told by the people connected to the instrument.

The tuner managed to restore the piano using as many of its original parts as possible and held a concert on the A-bomb anniversary in 2001. The reception was tremendous, and he felt it was his mission to convey the piano's origin as well as to make it sound the way it did before the bombing.

Since then, he has taken that first piano to more than 2,500 concert venues in Japan and abroad over the past 20 years, getting behind the wheel of a 4-ton truck himself to get the instrument to its domestic destinations. His activities have been reported nationwide, and he has now been entrusted with a total of six A-bombed pianos. In 2017, one of them was flown to Norway, where it was played at a concert to commemorate the Nobel Peace Prize awarded to the International Campaign to Abolish Nuclear Weapons (ICAN), an international NGO.

X

XIAO-MEI AND XIAN ZHANG

Chinese musician, pianist Zhu Xiao-Mei, in her book *The Secret Piano*, reveals much about the absence of pianos and Western music while growing up during the Cultural Revolution. She explains in her book that it was considered only 'bad' families would ever own a piano – such detestable objects of capitalist luxury. With some harrowing accounts in places, she wrote of music professors being publicly beaten and humiliated, and also being reduced to mere toilet cleaners at the universities where they had earlier been employed to teach music. Unsurprisingly, a few of them, deprived of being allowed to teach or even listen to music, even took their own lives. One, Gu Shengying, had performed at the very first concert Zhu Xiao-Mei ever attended: *It was rumoured that she had turned on the gas, sat down at her piano, and played Chopin's Funeral March.*

It was 1966, it had got to the stage when Zhu's family considered it just too dangerous to keep the piano at home any longer and decided it had to go:

My mother went out and flagged down the first Red Guards she met. She asked them to help her get rid of the piano immediately. They came in and took a look. "Out of the question," they said. "We're not touching it." A worthless thing, too heavy to move, was their assessment.

Throughout the book, the loyal little piano is obviously a very precious thing, almost a close member of the family. Despite being knocked around and having broken strings, although it had been sent out of the home and into hiding, it was reunited with its young owner, who was pleased to see it more or less in one piece.

Similar experiences were had by the Chinese conductor Xian Zhang (shown left). Known around the world, having conducted many orchestras, including at the Proms, she first learnt to play the piano on an instrument built by her father. She told one interviewer her first conductor's baton was a chopstick. As a youngster, she too recalled that under the Cultural Revolution, such things as Western music and pianos were banned.

Ironically, it could be said that in the Western world pianos in the home are not as widespread as they once were, yet the new cultural revolution is in China, where the piano is enormously popular. *The Economist* reported in 2019: *Of the 50 million children learning the piano worldwide, as many as 40 million may be Chinese. Shanghai alone has over 27,000 music schools.*

Y

YAYOI KUSAMA'S YELLOW-DOT PIANO; YAMAHA

For the Japanese artist Yayoi Kusama, polka dots have been part of her trademark for many decades; in fact, a London Underground tube map was designed using her familiar polka dots. In 2019 Tokyo's South Observation Desk in the Metropolitan Government Building – 202 metres high up on the 45th floor – was newly opened with the addition of a Yamaha grand piano (shown above) designed by Kusama. The Tocho Omoide (Memory) piano was put there having been inspired by other cities which have public pianos. From the 45th floor of Observation Desk can be seen views of skyscrapers and, on a good day, Mount Fuji. The piano is free for all to use, but perhaps following the Japanese culture of politeness and conformity, has a list of rules, the main one being a limitation of five minutes of playing per person.

Photo overleaf, on page 194: A government official at the opening ceremony.

Right: Can you spot the piano? Kusama is mainly based in America. This is another example of her work and is titled *The Obliteration Room*.

YAMAHA

The Yamaha company is quite an unassuming company, this is no doubt connected to the fact that the quality and reliability of their pianos speak for themselves. And while millions have been sold around the world to private owners and schools, it is the unassuming Yamaha that gets chosen time and again by professional musicians, major recording studios and international music competitions (demand is such that they are made in Japan, China, Taiwan and the USA). Used by countless famous names past and present, the Yamaha continues to be the people's and professionals' piano of choice.

Yet the company started slowly and steadily as an organ and reed manufacturer in 1887. The founder of Nippon Gakki Co., Ltd., Torakusu Yamaha, built his first upright piano in 1900. A grand piano followed soon after and the company went from strength to strength. It had sent a model as early as 1904 to the St Louis World's Fair, where it won an honorary Gold Prize, but jump to the 1960s and the company were using computer technology in their factories. Alongside modern technology, nonetheless, they had never been afraid to seek the best advice from craftsmen and pianists from Europe (going on to own Bösendorfer Pianos but allowing them to continue their traditional manufacturing process without interference). Yamaha launched the highly successful CF concert grand in 1969; the esteemed Russian pianist Sviatoslav Richter became a great champion of it after using one at a concert in Padua – appropriately enough the birthplace of the piano's inventor, Bartolomeo Cristofori. But of course, the wider company is also incredibly innovative and modern, making anything from motorbikes, digital keyboards, drums, vibraphones and sports goods to industrial robots...

Z

ZOOS; ZHU ZAIYU; ZEBRA

One doesn't normally associate zoos with pianos, but recently there have been zoos around the world which have allowed pianists to bring a piano along and play to the animals. Franklin Zoo in Boston, similarly, had a streetpiano set up close to where the zebras hang out so that they too could listen to anyone tinkling the (non-ivory!) ivories. Maybe one of the pianists was playing Michael Nyman's music for two pianos, *The Zoo Duet*. A bit before this time, London Zoo brought in a grand piano and called in the services of pianist Richard Clayderman to play romantic numbers for their rare tortoises – all in the hope that they would mate on Valentine's Day. Hopefully the Zoo found it worthwhile to shell out on the experiment.

ZHU ZAIYU

Perhaps he foresaw the arrival of the piano sometime after his death! The Chinese mathematician, physicist, musician (and much more) was ahead of the game when it came to tuning systems. It was he, Zhu Zaiyu (1536 – 1611, born in the Ming Dynasty), who described the equal temperament system, still used today, where an octave is mathematically divided into twelve semitones of the same interval.

ZEBRA (PIANO BARS)

Piano restaurants and bars in the UK have rapidly grown in popularity in recent years; seemingly, more punters are opting for real, live music played on acoustic pianos. In certain parts of America piano bars are something of an institution (not all are zebra-themed like Chicago's long-running Piano Lounge shown left!). As an example, Manhattan's Russian Samovar piano bar was reviewed by punchdrink.com:

An old-school joint of the sort that rarely survives in Manhattan these days, Russian Samovar has been around since 1986, when founder Roman Kaplan roped in Mikhail Baryshnikov and famed poet Joseph Brodsky as partners. Samovar is the kind of place where Liza Minnelli used to open her pipes after lapping up a bowl of borscht. Today, early evening diners eat as instrumental pianists do their thing, but after 8 p.m. – and especially on weekends – the scene becomes far more festive and interactive.

Most Requested Numbers: *Epitomizing Samovar's cross-continental appeal, Von Shats points to Frank Sinatra's rendition of 'New York, New York' and David Bowie and Mick Jagger's 'Dancing in the Street,' the latter of which she says is popular "especially for the line, 'Back in the USSR.'"*

Most Memorable Performance: *"The best ones are always spontaneous," says Von Shats, highlighting one instance when an opera company from Milan came straight from the Met and sang their hearts out until 4 a.m. "It was really amazing," she remembers. "After performing on one of the greatest stages in the world, coming in to dine, serenading our diners. It's just that kind of vibe."*

Above: In keeping with the A to Z theme of this book — here is an **A**merican **Z**ebra-themed grand from the Baldwin piano company, a brand still played and enjoyed in many an American piano bar and home today.

We end with two letter Zs in jazz, at least, and jump to France while maintaining an American theme. In 1911, famed Harry's Bar was literally dismantled brick by brick in Manhattan and rebuilt in Paris's Rue Daunou. It became an institution (it still is though more often with tourists walking the Opéra area of Paris). Much earlier, writers such as Sartre and Hemingway frequented it, and many a jazz legend has played in the underground piano bar, but its claim to fame ties in with George Gershwin, composer of *Rhapsody in Blue, I Got Rhythm, Swanee* and much more. As a twelve-year old boy, he surprised his parents by showing an interest in the piano his parents had bought for Ira, his older brother. George used it much more than Ira. George would later be turned down for coaching by the likes of composer Ravel, who recognized his talent lay in jazz rather than classical music. That Paris piano bar, to conclude, is where Gershwin sat down and composed his hit song *An American in Paris*. He wrote it on paper towels while there, but apparently during the war the keepsake towels were mistakenly used to light the stove. (Gershwin at one time owned three grand pianos.)

PICTURE CREDITS

Please note: Every effort has been made to attribute correct image with its owner and/or copyright holder, should an inadvertent omission or error have occurred these can be corrected in later editions by contacting the publisher.

Pages 2 to 99 p.2 Mrs I M Clark at the piano (1904): Meister Drucke, Cavendish Pianos; p.6 MEN/Bechstein; p.7 grand: Radford Piano Services, upright parts: infovisual; p.10 artnet; p.11 Abbey Road Studios: Reuters; p.12 Mills: Wikimedia, war factory: agfostock; p.13 Zeppelin: Wikimedia; p.14 Blüthner: Airships.net; p.15 Boldini: Mutual Art, Piano Builder, Dali: Tumblr, jazz poster: AllPosterCo; p.16 Van Gogh, Picasso, Munch: wikimedia; p.17 Vanity Fair, Liszt: Mary Evans Lib, Mozart Piano: artnet, Sentimental Colloquy: wikiart; p.18 Rebecca Horn: Tate Gallery, Sheffield Sculpture Park; p.19 Michael Parekowhai; p.20 Mozart Steinway: Robin Rile Fine Art, Paul Wyse Steinway: Steinway company, Carl Bechstein: Bechstein; p.22 Bechstein painting: akg-images.co.uk; p.23 Wagner piano: Josef Lehmkuhl, Bechstein Hall: Bechstein; p.24 Bechstein vehicle: Bechstein, Flat Spider: Steve Kerrccby, Thibaud: Wigmore Hall archives, poster: Mary Evans Lib; p.25 burning pianos: realizzazioniecatalogorostagno.it, Carlo Verdone; p.26 Elgar and shop: Elgar Society; p.27 Elgar's piano: National Trust; p.28 Holst Museum, Cole Porter Steinway: Elisa Rolle, Cole Porter and dog: NYTimes; p.29 Rachmaninoff: Wikimedia, Liszt Museum; p.30, 31 National Trust, Haydn portrait: Wikimedia; p.33 Dibdin: Holyrood Church; p.34 YouTube/Pathé News, Whaletone; p.35 Rolls Royce: loveisspeed.blog; p.36 Liberace pool: Liberace Museum, Disney pool: Walt Disney; p.39 piano frame: flickr; p.41 Finchcocks; p.42 modern Finchcocks: Garnot Keller, interior: Savills; p.44 Graf piano: Met. Museum of Art; p.45 Paderewski Centre; p.46 Paderewski: Britannica; p.47 Rubinstein: Mary Evans Lib; p.50/51 Rubinstein photos: Alfred A Knopf, statue: Wikimedia; p.51 Hess painting: Gallimaufry; p.52 Matthay: Pianosage.net; p.53 Horowitz: Wolffund.org; p.57 Horowitz: Boris Yurchenko, Wanda: Alchetron; p.58 Waterman and Brendel: Wikimedia; p.60 npr.org; p.62 Glenn Gould: Glenn Gould Foundation, Angelus News; p.63: Horowitz: NY Review, Brendel: Classic FM, Hess: Erica Stone, Rubinstein: Pinterest, Gould: IMZ International, Paderewski: Bain News Service; p.64 minipiano: Beatlesnews; p.65 Lennon: Redferns, Trident Studios, Hard Rock Hotel, White House Historical Association; p.66 expedition piano: Scott Polar Research Centre, Bonhams, NY Times; p.67 Atwell: Times Newspaper, Heath Steinway: David Cairns, Polish Embassy: Paderewskiego; p.68 piano art: Megan Brett, white Steinway: Nola Jazz Museum (USA), Priestley piano: V&A Museum, portrait: Sotheby's; p.69 Royal Suite: Claridge's Hotel, Alicia Keys piano: CBS News, The Beatles Museum; p.70 grand bookcase: Amazing Beautiful World.net, Wonderfuldiy.com, repurposed TV: desk-13; p.71 winerack: wonderfuldiy.com, roses: Rosen Tantau, waterfall: flickr.com, My Art magazine; p.72 blue: flickr.com, white: Designheroes.us, Liszt desk: Hartyányi Norbert; p.73 Guinness World Records; p.74 Adrian Mann; p.75 Sheffield Tabb, Grand Band; p.76 Piano Technicians' Guild, painting: Meister Drucke; p.77 Wikimedia, WordPress; p.78 Acabashi, Southern Lagniappe.blogspot, D. Moore:

Celebrity Graveland; p.79 Bechstein: Axel Mauruszat, Joyce: Wikimedia, Broadwood: Gatwick City Times; p.80 David Allen Barker, Schlieffen: Anne Sinclair/Diane Wats, Wikimedia, Schlitzberger and Daughters; p.81 The Piano Hotel, Claridge's Hotel; p.82 Hotel Grand Piano, Harrods: Matt Brown; p.83 Kensington and Chelsea Library; p.84 Selfridges Ltd, Gamages: Gumtree, London Transport Museum, British Newspapers; p.85 Pinterest, Elton John, Wikimedia; p.87 SmartMLS, Inc, ivory on wagon: Deep River Historical Society; p.88 Brigham Larson Pianos, Wikimedia; p.89 Metropolitan Museum of Art; p.91 Katherine Mansfield Society; p.92 USNews; p.93 Shutterstock; p.94 Skinner Auctions; p.95 Bath Piano Shop, Barton: Pastoral Project, joke by Jokejive.com, British Newspapers; p.96 Bach: News Hunter Magazine, Mozart: OnlinePianist, press any key: ART'm, Florida: unknown, bar tender: memegenerator.net; p.97 Amir Dotan, London Metropolitan Archives; p.98 Broadwood; p.99 Brinsmead: Mary Evans Lib, Guildhall Art Gallery.

Pages 100 to 200 p.100 PictureSheffield.com, Sheffield History.co.uk, Camberwell: author; p.101 photos by the author except Chappells: Aston Chase, Piccadilly: foursquare.com; p.102 Chernobyl: oddviser.com, Wikimedia, Steve Peck; p.103: National Library of Wales, war bomb site: Wikimedia, Player Piano Boat, East River: The Violin Channel; p.104 Lego/Piano Street, Lyre: SAP Renovation, Bösendorfer; p.105 Liszt: Amuraworld; p.106 Getty; p.107 Getty; p.112 TVTropes.org; p.114 ABC7NEWS, fallen grand: Iva Navatova, cat/concert: Baroque4You; p.115 British Newspapers, Broadwood; p.116 Graces Guide; p.117 Halifaxpeople.com, Bechstein, Graces Guide; p.118 Palings/Wikimedia; p.119 World Piano News; p.120 Scotsman Ltd; p.121 Wikimedia; p.123 The Pianist; p.124 Venn Pianos; p.125 upright: author, Venn Pianos, Peterson: Chicago Tribune; p.127 CTVNews; p.134 Architecture News; p.135 David Shankbone, Steinway Trail: Trip Adviser, Hunai: Unusualplaces.org, The Piano Café, Neumann/flickr; p.136 prepared pianos: Second Inversion; p.143 vrpianist.com; p.144 Country Life, Doll's Hse grand: Raphael Tuck & Sons, upright: Royal Collection Trust (RCT); p.145 The Piano Box: Metro Goldwyn Mayer; p.147 Adie/Piano Forum, Bösendorfer; p.148 Piano.org; Mary Evans Lib, short staircase: Paul Joseph, window: Rogers Removals; p149 Piano Workshop, Max Glover, helicopter: Ski Boutique; p.152 Broadwood; p.153 Royal Pavilion, Brighton, Pape square piano: RCT; p.154 invoice: RCT, Victoria and Albert: Getty, Erard: RCT; p.155 Bibliothèque Nationale de France, Bechstein: Luxurylaunches.com; p.156 RCT; p.158 YouTube, Wikimedia; p.159 Seesen Museum; p.160/1 Steinway; p.163/4 Atlasobscura.com; p.165 Tower Bridge: Chris Lukins, Leadenhall: Ayla and Sheryl/Streetpianos.com; p.166 Project Iguana, BBCNews, Coventry Telegraph; p.167 Canada: Michael McNamara; p.168 Herne Hill piano and St Pancras: author, Kiev: Sergey Dolzhenko; p.169 steps: Twitter@Brindille, Manilla: Reddit, New York: Brett Dahlberg, Playground: Live Journal, fountain playground: Astoria Post, Sweden: cmuse.org, interior staircase: grandado.com, Chile: c.muse.org; p.170 author's collection; p.171 BBC, b/w and white grand: Piano Builder, middle right: funophanticus.blogspot, top hat: Flickr, Root Simple; p.172 Nieuw: National Museums Northern Ireland, Canterbury Auctions Galleries BNP, The Luxury Train Club; p.173 The Piano Boat/Rhiana

Henderson and Masayuki Tayama; p.174 Weshare.hk; p.175 Uniplex (UK) Ltd, Worthpoint Auctions, Beethoven's tuning fork: British Library, wikimedia; p.178 H.W. Hatch: Alamy, Brinsmead: National Library of Wales; p.179 Perkins School for the Blind, White and Sentance; p.180 Christie's; p.182 Camping & Caravanning Club of GB, World Piano News; p.183 Pinterest, rectangular cake: Great British Bake Off.co.uk., Seabright Productions, Piano Forte: Amazon, Jeff Martin, Korg; p.184 stamp: Watford Observer, Sorrento: Alex Devil, US Naval Institute, upturned grand: UPM.com., Skytango.com; p.185 manchestersfinest.com, influenza: Wikipedia, Bösendorfer (p.186 also); p.188 Shackleford Pianos; p.189 Climate-ADAPT, Paris: Priscila Valentina; p.190 Gadzar/Piano Forum, researchgate.net; p.191 Blitz: Bridgeman, Wikicommons, Sophy Roberts, Reuters; p.192 La Guardia & Wagner archives; p.194 Chris Christodoulou/BBC; p.195 YK att-japan.net, Yamaha; p.196 Zebra Lounge; p.197 Baldwin Co; p.200 Soho Sq: English Heritage, Belfast Pianos.

Above: Joseph Kirkman premises at East Side, 3 Soho Square (looking south). The Kirkman firm (with different spellings over the years), renowned harpsichord makers, had even grander ambitions after newer family members joined the firm, diversifying into pianos.

Right: A typical modern showroom – Belfast Pianos – retailers of uprights and grands.

Still with us and going strong, the pianoforte can be both a humble and awesome musical instrument found in expected and many unexpected places right around the world...